Introducing groundwater

Michael Price

British Geological Survey

London
George Allen & Unwin
Boston   Sydney

George Allen & Unwin (Publishers) Ltd,
40 Museum Street, London WC1A 1LU, UK

George Allen & Unwin (Publishers) Ltd,
Park Lane, Hemel Hempstead, Herts HP2 4TE, UK

Allen & Unwin Inc.,
Fifty Cross Street, Winchester, Mass 01890, USA

George Allen & Unwin Australia Pty Ltd,
8 Napier Street, North Sydney, NSW 2060, Australia

First published in 1985

ISSN 0261-0531

British Library Cataloguing in Publication Data

Price, Michael
    Introducing groundwater.—(Special topics
in geology, ISSN 0261-0531; 2)
1. Water, Underground
I. Title   II. Series
551.49   GB1003.2
ISBN 0-04-553005-X
ISBN 0-04-553006-8 Pbk

Library of Congress Cataloging in Publication Data

Price, Michael, 1946-
    Introducing groundwater.
Bibliography: p.
Includes index.
1. Water, Underground.   I. Title.
GB1003.2.P75   1985   551.49   84-28455
ISBN 0-04-553005-X
ISBN 0-04-553006-8 (pbk.)

Set in 10 on 12 point Times by Mathematical Composition Setters Ltd,
Salisbury, UK and printed in Great Britain by Butler & Tanner Ltd,
Frome and London

This book is dedicated to the world's children

Preface

Although there are several excellent books on groundwater, it remains for many people a misunderstood and even mysterious substance. It seems to me that this is because all the textbooks on the subject are aimed at people who are, or intend to become, specialists in the subject; in effect the books are preaching to the converted. On the other hand, many of the more attractive introductory books on water mention groundwater only briefly. Whatever the reason, there is an unfortunate lack of understanding of the subject, even among geologists, geographers and civil engineers.

This book is an attempt to provide the non-specialist with a readable introduction to the subject. I have kept technical terms to a minimum, and where I have used them I have tried to explain why they are used. The emphasis is on principles rather than detail.

In a book of this kind little or nothing is original. However, I have assumed that readers will not want to see every factual statement supported by a reference; I have also assumed that anyone who wants more information will be likely to proceed initially to a more advanced book rather than to papers in academic journals. Therefore I have given references to books, and have referred only to those papers which are too recent to have been incorporated into books, or which are 'classics', worth looking at for that reason, and to those whose content has escaped mention in the existing books. To avoid breaking up the text, I have simply listed selected references at the end of each chapter.

The aim of the book is simplicity throughout, and I have kept mathematical formulae to a minimum. The entire text should be within the scope of anyone who has studied mathematics and physics or chemistry to GCE 'O' Level standard or equivalent. A knowledge of geology, though helpful, is not essential. I have relegated some of the more difficult concepts to self-contained boxes, which can be by-passed on first reading or by those who wish to avoid detail.

MICHAEL PRICE

Acknowledgements

Many people have contributed to this book by supplying help and advice. I am particularly grateful to my friends and colleagues at the British Geological Survey and in universities, colleges, and water authorities in the United Kingdom, who have not only helped directly in the writing of the book but who have added so much to my education over the years. During part of the time that I was working on the manuscript, I was fortunate enough to spend a year with the Geological Survey of Canada, and I should like to acknowledge the help and advice I received from friends at the GSC and elsewhere in Canada.

Specifically, I should like to thank John Barker for checking the accuracy of the physical statements, Adrian Bath for advising on water chemistry and Professor K. J. Ives for help with the public health aspects of water quality. Data and information on specialist topics were supplied by Southern Water Authority, Failing Supply Limited, and by Sigmund Pulsometer Pumps; I am particularly grateful to Graham Daw and Cementation Specialist Holdings Limited for advice on geotechnical problems. My special thanks are due to Dick Downing, whose seemingly inexhaustible knowledge of hydrogeology has been called upon throughout.

Early parts of the manuscript were reviewed by B. P. J. Williams and M. S. Money who provided useful criticism, and the entire manuscript has been reviewed by Dick Downing, Chris Wilson and Brian Knapp. Any faults that remain are the results of my stubbornness, not their omission. I am also grateful to Roger Jones and his colleagues at George Allen and Unwin for guiding the work to completion.

Finally, I am indebted to Mary, Susan and Ian for their patience and tolerance during my prolonged mental absences from family life!

MICHAEL PRICE

Contents

Preface   *page* vii

Acknowledgements   viii

1 Introduction   1

2 Water underground   3
   Selected references   12

3 Water in circulation   13
   Selected references   18

4 Caverns and capillaries   19
   Selected references   28

5 Soil water   29
   Selected references   36

6 Groundwater in motion   37
   Water movement   37
   Darcy's law   45
   Applications of Darcy's law   53
   Flow to a well   55
   Summary   63
   Selected references   64

7 More about aquifers   65
   Perched aquifers   65
   Confined aquifers: the concept and the misconceptions   66
   Elastic storage   69
   Fluctuations of water level   72
   Rock types as aquifers   77
   UK aquifers   79
   Other aquifers – unconsolidated sediments   84
   Non-aquifers   85
   Igneous and metamorphic rocks   86
   Selected references   87

8 Springs and rivers, deserts and droughts   88
      Discharge from aquifers   88
      Why rivers keep flowing   92
      Measuring river flows   94
      Hydrograph analysis   97
      What happens in a drought?   101
      Perennial drought – deserts   104
      Selected references   107

9 Water wells   108
      Pumps   109
      Well design   113
      Designing for maximum well efficiency   116
      Drilling methods   118
      Sampling and coring   123
      Selected references   125

10 Measurements and models   126
      Hydrological measurements   126
      Geological measurements   128
      Hydraulic measurements   132
      Models   148
      Water divining   151
      Selected references   152

11 Water quality   153
      Water for drinking   153
      Physical, chemical and biological aspects of quality 156
      Groundwater chemistry   156
      Chemical development of meteoric water   157
      Groundwater quality in arid areas   160
      Connate and saline waters   162
      Isotopes and tracers   163
      Groundwater temperature   167
      Selected references   168

12 Groundwater: friend or foe?   169
      Conjunctive use   170
      Groundwater as a problem   172
      Groundwater as a cause of instability   173
      Groundwater inflow to excavations   177
      Mine drainage   181

Selected references 182

13 Some current problems 183
Selected references 188

Index 189

1 Introduction

In the time that it takes you to read this sentence, three people will die because they do not have ready access to a safe and reliable supply of drinking water. According to figures recently issued by the World Health Organisation, an average of 50 000 people die each day from diseases associated with bad water; that is one person about every two seconds. The figures are a grim reminder of how much we need water.

This dependence is also illustrated by the way in which human settlements have grown up near reliable sources of water. It would be difficult to name any major town in the United Kingdom, or a capital city anywhere in the world, that does not lie on the banks of a river. There are admittedly other reasons for this – defence and communication being obvious ones – but the availability of water must always have been a key factor.

There are of course some areas of the world where, for reasons of climate or of geology, there are no permanent streams or rivers; yet many of these areas have been settled for thousands of years. The inhabitants of the early settlements relied for their supplies on water which occurs underground, often within a few metres of the surface, and which they exploited by digging wells. Sometimes the names of the settlements – names that in Britain end in '-well' or in the Middle East begin with 'Bir' or 'Beer' – testify to the nature of their water supply.

In the early cultures of Britain and the Middle East the origin of underground water may not have been understood, but its existence was known and exploited. In Britain today many people not only do not realise that much of the nation's water supply is drawn from underground, they are also unaware that underground water is common and widespread. As most people in Britain take their water supplies, along with their energy supplies, for granted, it is unlikely that they would stop to wonder where the water was coming from until the day when it failed to arrive; 'you never miss the water till the well runs dry' is a saying with a literal meaning in addition to the figurative one! The lack of awareness is perhaps inevitable, because the underground supplies are invisible and so cannot have the same impact on the senses as the expanse of a lake or the roar of a waterfall. It is however unfortunate, because underground supplies represent the largest accessible store of fresh water on Earth. They also frequently provide the best – in some cases the only – solution to the problem of providing water for drinking and irrigation in the Third World.

How does this underground water occur? How widespread is it, and how reliable as a source of supply? How can it be protected and best utilised for the benefit of mankind? A science called **hydrogeology** – a branch of geology devoted to the study of underground water – has grown up in an attempt to provide detailed answers to these and related questions. This book is an attempt to provide some of the answers in general terms for those people who are curious about one of our most precious assets – people who want to think about the water *before* the well runs dry.

2 Water underground

'In Xanadu did Kubla Khan
A stately pleasure-dome decree:
Where Alph, the sacred river, ran
Through caverns measureless to man
Down to a sunless sea.'
                              Samuel Taylor Coleridge

Many people share with Coleridge the idea that underground water occurs in vast lakes in caverns. They picture the water as flowing from one lake to another along underground rivers. Successful wells or boreholes, they imagine, are those which intersect these lakes or rivers; unsuccessful ones are those which encounter only 'solid' rock. The art of the water diviner or dowser is seen as predicting the location of these postulated subterranean watercourses, and so selecting a site for a borehole where water will be struck.

These popular misconceptions probably grew up because the only places where it is possible to see underground water in its natural state are the spectacular caverns which occur in hard limestones such as those of the Mendips and the Peak District. Visitors to these and similar caves are told that they were formed by the action of underground water. They hear of rivers that disappear into the ground in one place to reappear elsewhere, having flowed underground for part of their course. Not surprisingly, many of the visitors form the impression that this must be the normal mode of occurrence of all subsurface water. If this were so, since man is never slow to make money from his fellows, we might expect to find many other places where tourists could venture into the ground to see such features. The fact that they are relatively rare suggests either that underground water is restricted in its occurrence, or that it normally occupies less spectacular habitats. The latter possibility is borne out by the fact that in the regions where caverns are common, wells and boreholes are rare. Those parts of Britain that draw heavily on underground water for their supplies – such as the chalk lands of southern England and the sandstone tracts of the Midlands – contain no natural caves.

To understand the ways in which water occurs underground, we first need to think a little about the ground itself. Britain, like the rest of the Earth's crust, is made up of rocks of various types; in many parts of this country these rocks are not readily seen, because they are hidden by soil

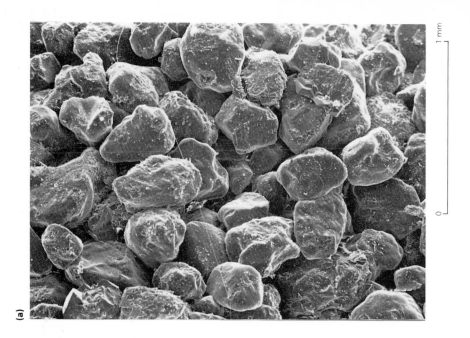

(a)

1 mm
0

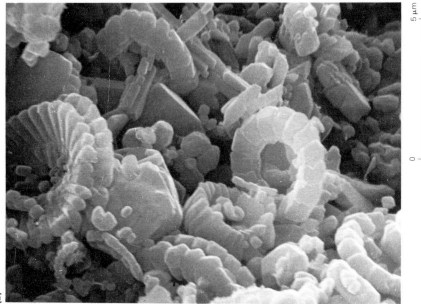

(b)

5 μm
0

(c)

(d)

Figure 2.1 Different types and sizes of voids in rock   (a) Photograph taken with an electron microscope (an electron photomicrograph) of sandstone of Permian age from a borehole in Cumbria. The porosity is 31 per cent. (b) Electron photomicrograph of chalk from a borehole in Berkshire, showing the minute fossils and shell fragments which make up the chalk. The porosity is 42 per cent but the small size of the pores means that in the absence of fissures the permeability is low. (c) Pore space resulting from fissures in limestone of Jurassic age (Great Oolite) in a quarry in the Cotswold Hills. Such fissures contribute most of the permeability of this aquifer. (d) A fissure which has been greatly enlarged by solution in Jurassic (Inferior Oolite) limestone, Cotswold Hills. ((a) and (b) are reproduced by permission of the Director, British Geological Survey.)

and the landscape is clothed by vegetation. But rocks can be seen in seaside cliffs, on mountainsides and, among other places, in quarries and cuttings. Wherever we live, if we probe deeply enough below the surface of the Earth, we shall eventually encounter rock.

In Chapter 7 we shall look in more detail at different types of rock, but for the present we need to note only one feature. This is that nearly all rocks in the upper part of the Earth's crust, whatever their type, age or origin, contain openings called **pores** or **voids**. These voids (Fig. 2.1) come in all shapes and sizes. Some of them − like the tiny pores in the chalk of the English downlands − are too small to be seen with the unaided eye. In exceptional cases, they may be tens of metres across, like the limestone caverns of Somerset and Derbyshire referred to earlier.

One common rock, sandstone, has pores that are more easily visualised (Fig. 2.1a). If you take a handful of sand and look at it closely − preferably through a magnifying glass − you will see that there are numerous tiny openings between the grains of sand. If you cannot readily find a handful of sand, look at granulated sugar which shows exactly the same feature. Sandstone is merely sand that has turned into rock because the grains of sand have become cemented together, and most of the openings will usually have been retained in the process − it is as though our granulated sugar had become a block of sugar.

The property of a rock of possessing pores or voids is called **porosity**. Rocks containing a relatively large proportion of void space are described as 'porous' or said to possess 'high porosity'. Soil also is porous. On a hot summer day the surface soil may appear quite dry, but if we dig down a little way the soil feels damp; if we could dig far enough to reach rock, this too would feel damp. The reason for this is that the pores are not all empty; some of them are filled, or partly filled, with water. In general it is the smaller pores which are full and the larger ones which are empty. At a still greater depth we should find that all the pores are completely filled with water, and we should describe the rock or soil as 'saturated'. In scientific terms we should have passed from the **unsaturated zone** to the **saturated zone**.

If we dig or drill a hole from the ground surface down into the saturated zone, water will flow from the rock into our hole until the water reaches a constant level. This will usually be at about the level below which all the pores in the rock are filled with water − in other words, the upper limit of the saturated zone. We call this level the **water table**.

The distance we need to drill or dig to reach the water table varies from place to place; it may be less than a metre, or more than a hundred metres. In general the water table is not flat; it rises and falls with the ground surface but in a subdued way, so that it is deeper beneath hills and shallower beneath valleys (Fig. 2.2). It may even coincide with the ground surface. If it does we can easily tell, because the ground will be wet and marshy or there will be a pond, spring or river. Where the water table is below the ground, as is usual, its depth can be measured in a well.

All water that occurs naturally below the Earth's surface is called **sub-surface water**, whether it occurs in the saturated or unsaturated zones. (I insert the word 'naturally' so as to exclude from the definition water in pipes, and the like.) Water in the saturated zone, that is to say below the water table, is called ground water or **groundwater**. Early practice was to write the term as two words but in Britain – and more recently in North America – there is an increasing tendency to write it as one word, to emphasise the fact that it is a technical term with a particular meaning. Needless to say some people dislike this; they argue that we should never write 'river water' or 'soda water' as one word, and so by analogy 'groundwater' should be 'ground water'. They doubtless have a point, but if we never allowed change in the English language we should still be writing as Chaucer did. So for the remainder of this discourse, groundwater it will be. If this offends the purists, they can mentally separate the term into two words each time it appears!

Why is groundwater important? One reason, mentioned in the introduction, is that in some areas it is the only source of water. A second reason is that groundwater represents a major proportion of the Earth's usable water resources.

It is estimated that the total amount of water on the Earth is about 1400 million cubic kilometres (km^3). Of this total, more than 97 per cent is sea water. About three-quarters of the remainder – in other words, about 2.2 per cent of the total – occurs in solid form in glaciers and polar ice-caps. Virtually all of the remaining water – the non-marine, unfrozen water – is groundwater. The water in rivers and lakes, in the atmosphere and in the unsaturated zone, together amounts to only about one-fiftieth of one per cent of the total world water supply.

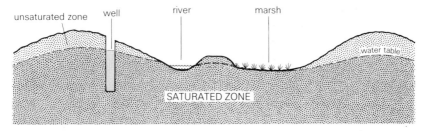

Figure 2.2 The water table

Groundwater thus accounts for about 8×10^6 km^3 (2 million cubic miles) or about 0.6 per cent of the Earth's total water. If we exclude ice, groundwater represents about 97 per cent of the non-marine water. About half of the groundwater occurs within 800 metres of the Earth's surface, and the remainder below this depth. Not all of it is usable; some for example is saline, and some can be regarded as inaccessible because of the great depths at which it occurs. Nor is it uniformly distributed over the land areas of the world. Where it does occur, it can be used – and has been used – to make deserts literally bloom; to make the difference between emptiness and plenty. (All estimates of the Earth's water resources are imprecise – estimating the volume of groundwater is particularly difficult, and values vary. The figure I have quoted is towards the lower end of the range, and can be regarded as a conservative estimate of the groundwater present in the Earth's crust.)

In terms simply of quantity, groundwater is therefore of great importance. But quantity is not everything, and groundwater has several other advantages over surface water as a source of supply. A surface reservoir must usually be impounded at one time, in one place, even though its full capacity may not be needed for many years. Groundwater can often be developed where and when it is needed, by sinking boreholes one at a time in appropriate places; if demand for water increases less rapidly than expected, the water-supply authority is not left with an expensive liability in the form of a man-made lake that nobody needs.

Unlike surface-water reservoirs which occupy large areas, frequently of prime agricultural land, the presence and utilisation of groundwater does not generally conflict with other use of the land under which it occurs. Deep beneath the ground it is unseen, insulated from changes in temperature, and protected from evaporation which, in a hot summer, can cause substantial losses of water from reservoirs and lakes. Its depth also renders groundwater less vulnerable to pollution, which is a potential threat to water on the surface.

The pollution risk to surface water is exemplified by an incident which occurred on 6 April 1978 in Yorkshire, when a barn containing stores of herbicides and pesticides caught fire. The chemicals, released when their containers were burnt or melted in the heat, were washed into drainage ditches by the water used in fire-fighting. The substances included a paraquat-based weedkiller, a deadly poison to which there is no known antidote. The contaminated water travelled along the ditches into the River Kyle, a tributary of the River Ouse, and so into the Ouse itself. The city of York draws most of its water supply from the Ouse.

On hearing the news of the incident, the York Water Works Company prudently closed its river intakes until the danger had passed. The Company's distribution reservoirs were sufficient to maintain the supply, but some consumers were inconvenienced as a result of the reduction in mains pressure. Had the danger not been immediately understood, or had the authorities been slower to respond, the consequences could have been serious. Although it would be foolish to imagine that groundwater is immune from such contamination, its natural cover of soil and rock provides it with considerable protection.

If groundwater has such advantages – abundance, availability in arid climates, and relative safety from pollution – one may be forgiven for wondering why we do not use it exclusively, or why we sacrifice large slabs of land in areas of outstanding natural beauty as reservoir sites. Surely there must be some drawbacks? There are.

Groundwater, despite its many advantages, has one serious disadvantage – it is not uniformly distributed throughout the Earth's crust. There are large areas of Britain, and of the world, where groundwater cannot be obtained in sufficient quantities to justify the expense of sinking wells and boreholes. At first sight this may seem to contradict what was said earlier, that nearly all rocks are to some extent porous and that below a certain depth – the water table – the pores are filled with groundwater. Since rocks are ubiquitous, it might seem reasonable to conclude that groundwater is available everywhere. It might seem reasonable, but it would be wrong.

Three main factors complicate the issue. The first is the extent to which the rocks are porous. If there are only a few small voids then the amount of water contained in a given volume of rock will be limited. So that we can make quantitative comparisons between different rocks, we define **porosity** as the ratio of the volume of the voids in the rock to the total volume of the rock. We usually express this ratio as a percentage, and so we speak for example of rocks with twenty per cent or thirty per cent porosity, meaning that the voids occupy respectively twenty or thirty per cent of the total rock volume.

In algebraic notation,

$$a = \frac{V_p}{V_b} \tag{2.1}$$

or

$$a = \frac{V_p}{V_b} \times 100 \text{ per cent}$$

where a = porosity, V_p = volume of voids, and V_b = bulk volume of rock.

The second factor is a combination of the size of the pores and the degree to which the pores are interconnected, because this combination will control the ease with which water can flow through the rock. We call this factor the **permeability**. Materials which allow water to pass through them easily are said to be **permeable**; those which permit water to pass only with difficulty, or not at all, are **impermeable**. A rock may be porous but relatively impermeable, either because the pores are not connected or because they are so small that water can be forced through them only with difficulty (Fig. 2.1b). Conversely, a rock which has no voids except for one or two open cracks will have a low porosity, and will be a poor store of water, but because water will be able to pass easily through the cracks the *permeability* will be high. Layers of rock sufficiently porous to store water *and* permeable enough to allow water to flow through them in economic quantities are called **aquifers**.

At great depths − typically 10 km or so below the Earth's surface − the rocks are so compressed and altered as a result of their deep burial that the voids have been closed and, for practical purposes, all the rocks are impermeable. These impermeable rocks form a floor below which groundwater cannot move − in effect they represent the bottom of the groundwater storage space available in the Earth's crust. Above this floor the storage space is filled, up to the level of the water table, with water that has entered the rocks − usually from rainfall.

In some areas, rocks of low permeability (such as clays or shales) may occur at or near the ground surface. None of these rocks are totally impermeable, but if they occur at the surface they will usually restrict the amount of rainfall that can soak into the ground. Similarly, if one of these relatively impermeable layers occurs *beneath* an aquifer, it will restrict the movement of water downwards from the aquifer. If part of the aquifer is also overlain by one of these layers, as in Figure 2.3, movement of water from the aquifer becomes so restricted that the groundwater in the aquifer becomes **confined** under pressure. In Figure 2.3, for example, rain water can enter the aquifer between X and Y. As a result of this supply of water the storage in the aquifer 'fills up' to give the water table in the position shown; the water table has to be above X, since this is the only place where water can flow from the aquifer. To the right of Z, the groundwater is confined between the impermeable layers.

The layers of low permeability are called **confining layers** or **confining beds**: the aquifer is called a **confined aquifer**. It is important to note, however, that between X and Z the aquifer is **unconfined**. To the right of Z, where the aquifer is confined, it has no water table and no unsaturated zone − the permeable material is saturated for its full thickness. If a well is dug or drilled into the upper confining layer (for example, well C in Fig. 2.3), it will encounter no groundwater (except

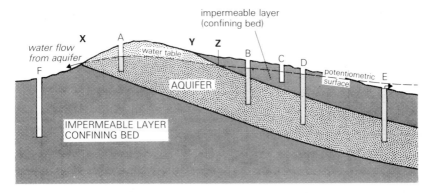

Figure 2.3 A confined aquifer and its potentiometric surface Between X and Z the aquifer is unconfined and has a water table; to the right of Z the aquifer is confined and has a potentiometric surface. Wells A, B, D and E enter permeable material and strike water. Wells C and F are in impermeable material which will yield water only very slowly.

perhaps a very slow seepage from the 'impermeable' material); if the well is deep enough to reach the aquifer (as at B and D in Fig. 2.3) the water will rise up in the well because the water in the aquifer is under pressure. The level at which the water stands in the wells defines an imaginary surface whose height above the aquifer depends on the pressure in the aquifer; this surface is called the **potentiometric surface**. Sometimes the potentiometric surface may rise above ground level, in which case the well (like E in Fig. 2.3) will overflow.

We shall talk more about confined and unconfined aquifers in Chapter 7, but for now we have seen something of the influence of permeability on the availability of groundwater.

The third factor that determines the amount of groundwater available from the rocks of an area is the amount of replenishment – the degree to which water abstracted from the aquifer is replaced. The replenishment may come from above, as a result of rainfall soaking into the ground, or it may take place laterally or from below, from adjacent aquifers carrying water from elsewhere. The replenishment factor depends not only on the nature of the rocks but on the soil and vegetation which cover them and on the climate of the region. It is part of the **water balance** of the area – the balance between the water which enters the area and the water which is used or which leaves it. In assessing the groundwater resources of any region, knowledge of the water balance is as vital as knowledge of the porosity and permeability of the rocks. This is because groundwater is not isolated from other water; as we have seen, it is part of the Earth's total store of water. As such, it is in more or less continuous interchange with all other water in a system of circulation called the **water cycle**, which is the subject of the next chapter.

Selected references

Leopold, L. B. 1974. *Water; a primer*. San Francisco: W. H. Freeman.
Nace, R. L. 1969. World water inventory and control. Chapter 2 of *Water, earth and man* (ed. R. J. Chorley). London: Methuen.
Pearce, F. 1981. Water, water everywhere. . . . In *New Scientist* **92**, no. 1274 (8 October 1981), 90–3. (Discusses the recent decisions on the building of reservoirs in England and Wales.)

3 Water in circulation

It is a matter of observation that virtually all rivers discharge their waters into the sea. As the level of the sea remains more or less constant, and as the rivers show no apparent sign of ceasing to flow, there must be some mechanism by which water is returned from the sea to the land at the same average rate at which it flows via the rivers to the sea. The nature of this return mechanism is less obvious than the clearly visible flow of the rivers, and has occasioned much speculation throughout history.

We now know that the return path is through the atmosphere. The energy of the Sun's rays causes water to evaporate from the surface of the oceans (Fig. 3.1). The water vapour so produced is a normal part of the Earth's atmosphere; it remains in the atmosphere, completely invisible, unless cooled sufficiently to cause it to condense and form water droplets. The cooling occurs as a result of the vapour-laden air rising into higher and colder regions of the atmosphere. This may result from the air being forced upwards over a physical barrier, such as a mountain range, or over a meteorological barrier, such as a mass of colder, denser air; or it may result from convection. Whatever the cause, the chilled vapour condenses as droplets which form around suitable nuclei (tiny

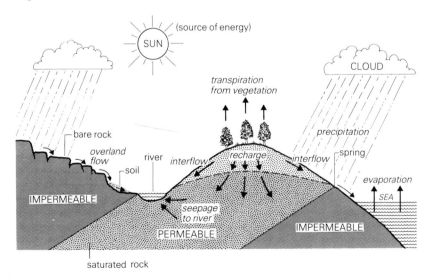

Figure 3.1 Diagrammatic representation of the water cycle

dust and other particles) which are nearly always present in the atmosphere. When the droplets are present in sufficient quantity, we see them as clouds.

These water droplets are much smaller than raindrops, and are light enough to remain suspended in the atmosphere. The process by which they grow until they are large enough to fall as rain is surprisingly complicated. To me, living in a temperate, humid region, where rain is generally regarded as a persistent nuisance until we have been without it for a few weeks, it came as something of a shock to learn that it actually needs a favourable combination of circumstances to produce rain, but such is the case.

A typical thundercloud contains more than 50 000 tonnes of water. At any one time, the atmosphere contains about 14 000 km^3 or 14×10^{12} tonnes of water. These may sound enormous amounts, but the total atmospheric water is in reality about one-thousandth of one per cent of the Earth's water, or 0.03 per cent of non-oceanic water. If all this atmospheric water were simultaneously to fall as rain, it would produce an average of about 27 mm of rainfall over the total surface of the Earth. The average annual precipitation over the Earth is believed to be about 900 mm, which means that the equivalent of about 33 total condensations of the atmospheric water occur each year. A large amount of precipitation must occur over the oceans, thereby short-circuiting the water cycle.

Water that is precipitated over land areas may take several routes through the remainder of the cycle. Some of it will never reach the ground surface. It will be **intercepted** by foliage, and held on the leaves of trees and plants until it evaporates. Some of the water reaching the ground will fall on bare impermeable rock or on artificial, paved surfaces. Apart from that which collects in depressions and remains there until it evaporates, this water will run off these surfaces into natural or artificial drainage channels. The remainder of the water will fall on soil, and it is largely the condition of that soil which will determine what happens to the water thereafter. Broadly speaking, the water falling on soil may be disposed of in three ways: it may be evaporated, either directly or by transpiration from vegetation after being drawn up by plant roots; it may run over the surface of the soil or travel in the near-surface soil layers until it reaches a ditch or stream; or it may soak into the deeper layers of the soil and so perhaps into the underlying rock. In the early part of this century, when these processes were understood even less well than they are today, it was commonly assumed that the rainfall over a particular area would always be disposed of in more or less the same proportions. It was common to hear engineers talk of the evaporation fraction, the runoff fraction, and the percolation fraction. Now that we understand these things more fully – or at least, think we do – we know that the division of precipitation into these end products is more complex.

When rain begins to fall on relatively dry soil, we know from experience that it is readily absorbed. In everyday terms we say that the rain soaks into the ground; in scientific terms we speak of **infiltration**, which is the process whereby water enters the ground at the surface.

If the soil is dry and the rainfall is light, all the water reaching the ground will infiltrate into the soil and be held there as films of moisture which surround the individual soil particles. This water is held in the soil until it is either evaporated directly from the soil surface or taken up by the roots of plants. A small fraction of the water which the roots take up is retained in the plants as part of their growth process, but the majority is evaporated from openings in the leaves and stems in the process known as **transpiration**. The combined effects of evaporation and transpiration in returning water to the atmosphere are frequently grouped together and termed **evapotranspiration**.

As each successive layer of the soil absorbs water, infiltration moves on downward through the soil and subsoil to the underlying rock. If this rock is permeable, the infiltration process will continue downwards, through the unsaturated zone, until the infiltrating water arrives at the water table and joins the groundwater in the saturated zone. Precipitation reaching the water table is called **recharge**, because it is helping to replenish the store of groundwater.

The maximum rate at which water can enter the soil is called the **infiltration capacity** of the soil. If the rainfall is exceptionally heavy, a situation may arise where water is arriving at the surface of the soil more quickly than it can enter the soil; in this case, the infiltration capacity is said to have been exceeded by the rate of rainfall. In these circumstances, the soil will behave like an impermeable surface; depressions will fill, and then **overland flow** (water flowing across the ground surface, usually as small trickles and rivulets) will occur. **Surface runoff** is that part of **total runoff** (the river flow leaving the area) which results from overland flow. This overland flow is sometimes called **Hortonian flow** after R. E. Horton, the American hydrologist who put forward the theory of infiltration capacity.

In practice, this type of overland flow is rare. In areas with well developed soils, it occurs only when the soil is frozen or during exceptionally heavy rain. Even in places where vegetation is sparse or where soils are thin or absent, it seems that what begins as overland flow often soaks into soil elsewhere before it can reach a drainage channel as surface runoff. Modern theories actually attribute little or no stream flow to surface runoff, placing much more emphasis on subsurface flow. However, not all the water that has infiltrated the soil reaches the water table. If the rocks beneath the soil are impermeable, or if there are layers in the soil which are themselves of contrasting permeability, there will be a tendency for water to move laterally through the unsaturated zone until it eventually arrives at stream or river channels. This flow is called **interflow**, because it is intermediate between overland flow and true groundwater flow.

Water that has reached the water table becomes groundwater. It **percolates** slowly through the aquifers, at rates which under natural conditions may vary from more than a metre in a day to only a few millimetres in a year. The groundwater moves towards an outlet from the aquifer, which is usually a point where the water table intersects the ground surface. Where this occurs, water will seep or flow from the aquifer; in doing so, it will cease to be groundwater and will revert to being surface water, usually finding its way into a river channel.

The water in rivers usually finds its way back to that primary store of the Earth's water, the sea, but there are some exceptions. Some water will evaporate from the river surface and return to the atmosphere. Under certain conditions, water may flow from the river into the ground, so that rivers may lose water to an aquifer as well as gain from it. There are some parts of the world, such as intermontane basins in semi-arid regions, where the combination of these effects is sufficient to cause rivers to dry up completely, long before they can reach the sea. These are the exceptions, however, to the general Biblical rule that 'All the rivers run into the sea' (Ecclesiastes 1, v. 7), thereby completing the water cycle.

The water cycle is also referred to as the **hydrological cycle**. Hydrology is 'the science that treats of waters of the Earth, their occurrence, circulation, distribution, their chemical and physical properties, and their reaction with their environment, including their relation with living things'. This definition was prepared for the President of the United States in 1962, but hydrology is much older than this implies. We know that man has speculated on the origins of rivers, and on what sustains their flow, since the earliest civilisations. Today, our version of the hydrological cycle seems so logical and obvious that it is difficult to believe that it did not gain widespread acceptance until the 16th century. This was caused in large part by the tendency of the philosophers of Ancient Greece to distrust observations and by the tendency of later philosophers to accept the opinions of the Greeks almost without question. Plato advocated the search for truth by reasoning. He and his followers appear to have attached little importance to observations and measurements. Thus Aristotle, Plato's most famous pupil, was reportedly able to teach that men have more teeth than women, when simple observation would have dispelled this idea. From a hydrological viewpoint, however, he had a more serious misconception – he believed that rainfall alone was inadequate to sustain the flow of rivers.

This error could not be corrected until it was realised that observation and measurement are an essential part of the advancement of scientific knowledge. The first person to make a forthright and unequivocal statement that rivers and springs originate entirely from rainfall appears to have been a Frenchman called Bernard Palissy, who put forward this proposition in 1580. Despite this, in the early 17th century many workers were still in essence following the Greeks in believing that sea water was drawn into vast caverns in the interior of the Earth, and raised up to the level of the mountains by fanciful processes usually involving evaporation and condensation. The water was then released through crevices in the rocks to flow into the rivers and so back to the sea.

In 1674 Pierre Perrault, a French lawyer, published anonymously a book on the origin of springs in which he described how he had compared the annual flow of the headwaters of the River Seine with the amount of precipitation falling each year on its catchment. He found that the precipitation was equal to about six times the flow, and concluded that, in general, 'the waters of rains and snows are sufficient to cause the flow of all the Rivers of the World'.

Edmé Mariotte, another Frenchman and a scientist of great standing, carried out a similar exercise for the whole of the Seine catchment above Paris. He arrived at a similar conclusion, though his work was not published until 1686, two years after his death.

If Perrault and Mariotte demonstrated that precipitation could supply all the flow of the world's rivers, it fell to an Englishman to demonstrate the logic of the other half of the hydrological cycle – the ability of evaporation from the oceans to account for the precipitation. Edmund Halley is best known as an astronomer and as a friend and colleague of Isaac Newton. During astronomical observations from the island of St Helena, Halley was troubled by condensation on the lenses of his telescopes and became so interested that, on his return to London, he carried out a series of experiments. As a result of calculations based on measurements of the rate of evaporation from a pan of water, Halley felt able to conclude in 1693 that the water which evaporates from the Mediterranean Sea on a typical summer day would be three times the amount flowing into the sea on that day in the major rivers. In other words, in general terms, there is sufficient evaporation from the seas to account for river flow.

Perrault, Mariotte and Halley may be described as the first quantitative hydrologists. They established the concept of the hydrological cycle not by speculation, but by observation, measurement and calculation. Groundwater is part of that cycle; since hydrogeology is the study of groundwater, it follows that hydrogeology is a subdivision of hydrology as much as it is of geology. So it is: just as water in its natural cycle takes different forms in different realms – ice, liquid and vapour in glacier, sea, river, aquifer and atmosphere – so hydrology encompasses many fields of study. It is a multi-disciplinary science, involving mathematics, physics, chemistry, meteorology, geology and glaciology, in addition to biology in all its diversity. Hydrogeology must be just as broadly based because, as later chapters will show, it is impossible to isolate completely one portion of the hydrological cycle from the others.

Selected references

Biswas, A. K. 1970. *History of hydrology*. Amsterdam: North-Holland.
Mason, B. J. 1975. *Clouds, rain and rainmaking*, 2nd edn. Cambridge: Cambridge University Press.
Nace, R. L. 1969. World water inventory and control. Chapter 2 of *Water, earth and man* (ed. R. J. Chorley). London: Methuen.
Todd, D. K. 1980. *Groundwater hydrology*, 2nd edn. New York: Wiley.
Vallentine, H. R. 1967. *Water in the service of man*. London: Penguin.
Ward, R. C. 1975. *Principles of hydrology*, 2nd edn. Maidenhead: McGraw-Hill.

4 Caverns and capillaries

Water is a commonplace and probably, to many people, a rather uninspiring substance. Yet it has a number of unusual properties which not only influence its behaviour underground but which also have a great effect on our lives, and which combine to make it unique.

To begin with it occurs on Earth, under natural conditions, in all three physical states — as a solid, as a liquid and as a gas — a fact that is true of no other common substance. When it freezes, instead of contracting like other substances, it expands. This expansion is a nuisance to the householder and to the water engineer, because it can cause water pipes to burst; to the geologist it means that water at about freezing point is a powerful geological agent, because the freezing of water in rock crevices can eventually shatter rocks into fragments.

Another thermal oddity is water's high specific heat — it takes more heat to raise the temperature of a kilogramme of water by one degree than it takes for any other common substance; similarly, water gives up more heat on cooling. This is one reason why the sea has such a stabilising effect on the climate of adjacent land areas, and why water is so effective in putting out fires.

Water is also one of the most effective solvents known. There are few materials that do not dissolve in it to some extent; each time you drink a glass of water, for example, the water will have a small amount of glass dissolved in it.

Water containing dissolved carbon dioxide is capable of dissolving calcium carbonate, the main constituent of limestone, which is only slightly soluble in pure water. Water falling as rain dissolves some carbon dioxide on its passage through the atmosphere, and much more when it infiltrates the soil. When rain falls on a region composed of limestone, it will infiltrate and percolate through cracks and crevices, slowly dissolving the rock. If there is no exit for the water, it will become saturated with calcium carbonate and no further dissolution will occur, but if a throughflow becomes established the dissolved material will be carried away and the crevices enlarged by further dissolution. In this way the giant caverns of Somerset, Derbyshire and elsewhere in Britain have been formed; there are more impressive examples overseas, like the Mammoth Cave System of Kentucky and the Carlsbad Cavern of New Mexico, which has one chamber more than a kilometre long, 200 m wide and almost 100 m high.

The dissolution of calcium carbonate is a reversible process (the material is actually transported in the form of the bicarbonate, the chemical equation being $H_2O + CO_2 + CaCO_3 \rightleftharpoons Ca(HCO_3)_2$). As the water with its dissolved material moves through crevices and caverns, then evaporation, partial loss of carbon dioxide, or the change in pressure may result in calcium carbonate being precipitated, especially if the cave is only partly filled with water. If precipitation occurs as water drips from the roof of a cave, the result will be a **stalactite** – an 'icicle' of calcium carbonate – growing slowly downwards from the roof. Water dripping from the stalactite may give rise to further carbonate precipitation on the cave floor, so that another structure – a **stalagmite** – will grow upwards. (An infallible way of remembering which way stalactites and stalagmites grow is to think of 'ants in your pants' – 'mites' go up, 'tites' go down!) The two may eventually join, forming a continuous column from floor to ceiling. Clusters of stalactites and stalagmites may take on strange and beautiful forms and are often given fanciful names, especially in caves which are easily accessible to visitors.

Where water runs in sheets down cavern walls, the deposits themselves become curtain-like structures of calcium carbonate in a form called **travertine**, and may look like solid waterfalls. Objects placed in the path of this carbonate-rich water will become covered with calcium carbonate. This phenomenon has led to the advertisement of 'petrifying wells', where one may see everyday objects that have become coated in this way. These have not been petrified in the geological sense, which would mean that each molecule of the original material had been replaced with a molecule of a mineral, but have merely become encrusted.

However, as was said earlier, these limestone caverns and their beautiful structures are a rarity. In most limestone areas groundwater moves through crevices or fissures which, even when enlarged by dissolution, are rarely more than a few millimetres across and which are often less than a millimetre. In other rock types, such as sandstones, it is rare for fissures to be enlarged by dissolution at all, and the permeability and the flow of water are much more diffuse.

In these granular materials another property of water called **surface tension** comes into play. This is not unique to water but is common to all liquids; it arises from the attraction that the molecules of the liquid have for each other, and has the effect of making the liquid surface behave almost as though it were an elastic membrane. In reality of course there is no membrane, but many of the effects that occur at a liquid boundary can conveniently be considered to be the result of a force – surface tension – acting in all directions parallel to the liquid surface.

Surface tension has units of force divided by length, so in the SI system it is expressed in newtons per metre (N/m).

One of the effects of this molecular attraction or surface tension is a tendency to reduce the free surface area of any body of liquid to a minimum. The geometrical shape that has the smallest surface area for a given volume is a sphere, so we should expect a mass of water, freely suspended, to assume a spherical shape. We see in drips from taps and in raindrops that, subject to modification by gravity, this is so.

When pure water is spilled on clean glass, the water spreads over the glass in a thin film, tending to wet the glass evenly. This is because the attraction between the molecules of water and the molecules of glass is *greater* than that which molecules of water have for each other; as a result, the water spreads itself over the glass in a layer which, in the limiting case, would be only one molecule thick. Because in this case the surface of the water and the surface of the glass are parallel to each other, we say that the **angle of contact** between the surfaces is zero (Fig. 4.1a).

If the sheet of glass is dirty the water does not spread out so evenly; we say that it has less tendency to 'wet' the surface. It no longer tends to form a uniform film, but instead forms discrete globules (Fig. 4.1b), each of which has a finite angle of contact θ with the glass surface.

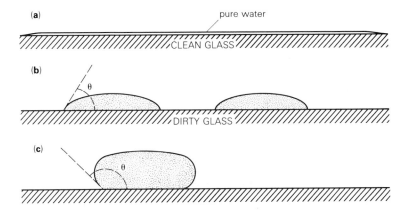

Figure 4.1 Surface tension (a) On clean glass, pure water spreads out to form a thin film. (b) On dirty glass, water shows less tendency to wet the whole of the surface. It tends to stay in globules, with a finite angle of contact θ between the water surface and the glass. (c) In the case of mercury on glass, or water on wax, the tendency to form globules is very pronounced and the angle of contact θ is greater than $90°$. For mercury on glass, θ is about $140°$.

In some cases the molecules of a liquid have a greater attraction for each other than they have for the molecules of the solid with which they are in contact. In these cases the solid appears to repel the liquid; the angle of contact between liquid and solid is greater than 90 degrees, and we say that the liquid is 'non-wetting' to that solid (Fig. 4.1c). Examples are mercury on glass, and water on wax. The reason why wax polish helps to protect cars is that water is repelled by the wax, and rolls from the surface more easily than it would from the un-waxed paintwork.

A result of surface tension that is of particular importance to us is the phenomenon called **capillarity**. If a glass capillary – a tube with a fine bore – is held more or less vertically with its lower end dipped in water, the water rises up the capillary tube by an amount that depends on the radius r of the tube (Fig. 4.2a). The surface of the water comes to rest at a vertical height h above the free surface of the water outside the tube. Also, the water surface inside the capillary is not flat; close examination shows that, like any liquid surface, it is curved where it comes into contact with the solid walls of the tube. We recognise this as a direct result of surface tension; once again the attraction of the glass molecules for the water molecules is greater than that of the water molecules for each other, so that the water is pulled upwards around the circumference of

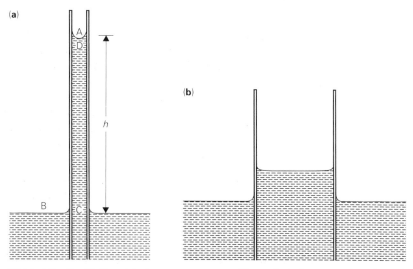

Figure 4.2 Capillarity (a) In a narrow capillary the air–water interface is hemispherical. The pressure at D is less than the pressure at A and therefore less than the pressure at B, which, like A, is at atmospheric pressure. Water rises up the tube until its hydrostatic pressure compensates for this pressure reduction and the pressures at B and C are the same. (b) In a wide capillary, the interface is not a perfect hemisphere; the central portion is flattened.

the tube to form the shape we call a **meniscus**. In a wide tube the meniscus will have a flattened central area (Fig. 4.2b), but in a narrow, circular tube the meniscus is shaped like a segment of a sphere; if the angle of contact between water and glass is zero, then the radius of this sphere will be the same as the radius r of the capillary tube, and the segment will be a hemisphere.

When the interface between a liquid and a gas is curved, the pressure on the concave side of the interface is greater than that on the convex side. In Figure 4.2a the pressure at A and the pressure at B are equal (they are both atmospheric), and when the water column is at equilibrium the pressure at C must equal the pressure at B. Because A is on the concave side of the meniscus, the pressure at D is less than that at A and B; therefore the height of the water column must be such that the extra pressure which it causes at C is compensated for by the pressure drop across the meniscus. In practice, for pure water in a clean glass tube, this occurs when

$$h = \frac{15}{r} \qquad (4.1)$$

with h and r both measured in millimetres. A little thought shows that if the water at B and C is at atmospheric pressure, then at all points between C and D its pressure is less than atmospheric.

All this may seem a little remote from hydrogeology. Surely rocks do not have long thin cylindrical tubes running through them? They do not, but the interconnected pores in a granular material can behave as though they were bundles of capillary tubes. Let's go back to our substitute sandstone, the lump of sugar. If a lump of sugar is held so that its lower surface is in contact with tea or coffee, the liquid will be seen to be drawn up into the pore spaces in the sugar lump. We cannot observe this for long because the sugar soon begins to dissolve, but the experiment demonstrates that capillary effects do occur in porous materials.

We would therefore expect to find surface tension and capillary phenomena occurring in rocks, and we should not be disappointed. One of the best known is the **capillary fringe** (Fig. 4.3). This is a layer of rock immediately above the water table in which water is held by capillarity. Within the capillary fringe the rock is still more or less saturated, but it is distinguishable from the saturated zone proper by the fact that a well will fill with water only to the base of the capillary fringe, i.e. to the water table. Water in the capillary fringe is referred to, not unreasonably, as **capillary water**, to distinguish it from the true groundwater of the saturated zone.

The physical difference between groundwater and capillary water,

The height of the water column in a capillary

In general terms, the pressure drop across a spherical gas/liquid interface is equal to $2\sigma/R$, where R is the radius of curvature of the interface and σ is the surface tension of the liquid. In the case of capillary rise, the pressure deficiency caused by the curvature of the meniscus (Fig. 4.2a) must equal the pressure at C caused by the height h of the liquid column; this pressure is $\varrho g h$, where ϱ is the liquid density and g is gravitational acceleration. Hence

$$\frac{2\sigma}{R} = \varrho g h$$

so that

$$h = \frac{2\sigma}{R\varrho g}. \tag{4.2}$$

If the angle of contact is θ, then $R = r/\cos\theta$ so that

$$h = \frac{2\sigma\cos\theta}{r\varrho g}. \tag{4.3}$$

For pure water in clean glass, θ is approximately zero, so that $\cos\theta = 1$ and $R = r$, i.e. the meniscus is a hemisphere with the same radius as the capillary tube. At $10\,°C$, $\sigma = 0.074$ newton/metre and ϱ is approximately 1000 kg/m^3. Therefore, taking g as 9.81 m/s^2,

$$h = \frac{2 \times 0.074 \times 1}{r \times 10^3 \times 9.81} = \frac{1.5 \times 10^{-5}}{r}$$

with h and r in metres, or

$$h = \frac{15}{r}$$

with h and r in millimetres.

which explains why wells fill to the level of the water table, can be seen if we refer back to Figure 4.2a. We said that the water between C and D was at a pressure below atmospheric, while that at B and C was at atmospheric pressure. In the ground, water in the capillary fringe (like water in the capillary tube between C and D) is at *less* than atmospheric pressure; below the water table the groundwater is at a pressure *above* atmospheric, in accordance with the laws of hydrostatics. This is illustrated in Figure 4.4. This gives us a way of defining the water table that is more general – and therefore more useful – than simple reference to the level at which water stands in a well. We can say that the **water table** is that surface in an underground water body at which the water pressure is exactly equal to atmospheric pressure.

The thickness of the capillary fringe depends on the effective radii of the capillary tubes formed by the interconnected pores of the rock formation – in other words, on the pore sizes. It thus varies from one rock type to another. In a coarse gravel it may be only a few millimetres thick; in chalk or clay it may be several metres. The pores of the material will not be all the same size, so the top of the capillary fringe will not be an abrupt surface but an irregular and gradual transition.

If the water table in an aquifer falls for any reason, the capillary fringe falls with it. However, the rock that was formerly part of the capillary fringe does not drain completely of water; surface tension and molecular effects cause a thin film of water to stay in place around each particle of rock material (Fig. 4.4). This is important, because it means that not all the water in the pore space of an aquifer can be abstracted and utilised. If the water table over an area A of an aquifer falls by an amount z, then the volume of rock affected is $A \times z$. The pore space contained within this volume is $A \times z \times a$, where a is the porosity expressed as a fraction. But the volume of water that will drain from the aquifer is $A \times z \times S_y$, where S_y is less than a and is called the **specific yield** of the aquifer. Specific yield is the ratio of the volume (V_w) of water that will drain by

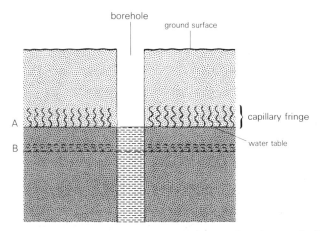

Figure 4.3   The capillary fringe   If the water table is lowered (e.g. by pumping) from A to B, the capillary fringe will also be lowered. If the rock properties at B differ from those at A, then the new capillary fringe may be thicker, or – as in this case – thinner than the original one.

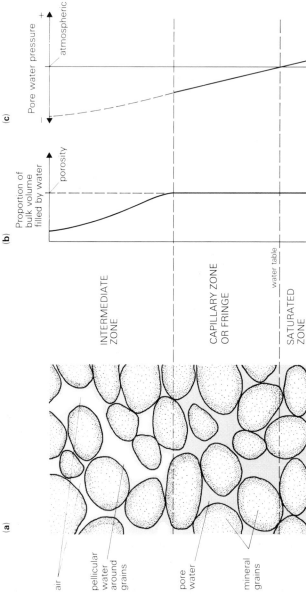

(a)

air

pellicular
water
around
grains

pore
water

mineral
grains

(b)

Proportion of
bulk volume
filled by water

porosity

(c)

Pore water pressure

atmospheric

+

−

INTERMEDIATE
ZONE

CAPILLARY ZONE
OR FRINGE

water table

SATURATED
ZONE

Figure 4.4 Pore water in the unsaturated zone (a) An exaggerated view of the intermediate zone, with pellicular (film) water around mineral grains, above a capillary zone and water table. (b, c) Graphs of pore-water content, (b) and pore-water pressure (c) for the section shown in (a). The precise shapes of the upper parts of the curves in (b) and (c) will depend on previous infiltration and evaporation events. It is important to realise that (a) is a two-dimensional section through a three-dimensional aggregate of grains – this is why in the figure it appears that not all grains are in contact with their neighbours. There will be many air-filled pathways through the intermediate zone.

gravity from a rock or soil that was initially saturated to the volume (V_b) of the rock or soil, i.e.

$$S_y = \frac{V_w}{V_b} \tag{4.4}$$

or

$$S_y = \frac{V_w}{V_b} \times 100 \text{ per cent.}$$

(Strictly speaking, all the water that drains from the aquifer in the example above does not come from the volume of rock $A \times z$. Some of the water will have drained from the original capillary fringe above this volume, while some of the water within the volume under consideration will be retained in a 'new' capillary fringe. If the rock properties vary considerably between the two positions of the capillary fringe, this fact may have an important and complicated effect upon the drainage behaviour. See Figure 4.3.)

The water that is unable to drain from the pores is referred to as **specific retention** (S_r), which is the ratio of the volume (V_r) of water that a rock or soil, after being saturated, will retain against the pull of gravity, to the volume (V_b) of the rock or soil, i.e.

$$S_r = \frac{V_r}{V_b}. \tag{4.5}$$

The definitions of specific yield and specific retention assume that gravity drainage is complete when the ratios are measured.

It is obvious that the sum of specific yield and specific retention is porosity. In coarse-grained rocks with large pores, the capillary films occupy only a small proportion of the pore space and the specific yield will almost equal the porosity. In fine-grained rocks like chalk, capillary effects are dominant and specific yield will be almost zero.

If you have ever laundered a large article like a blanket or bath towel, you will have experienced one of the effects of surface tension on drainage. You can take a large, wet towel and squeeze out as much water as possible, but within a short time of hanging it on a clothes line the lower edge will be saturated and dripping water. The reason for this is that while the towel is being squeezed or wrung, the capillaries are relatively short, so that their effective length is *less than* that of the column of water which they can support; consequently, they remain full of water. When the fabric is hanging vertically, the capillaries are *longer than* the water column which they can support, so water begins to drain from the bottom of them, and so from the bottom edge of the fabric.

This means that if we take a lump of rock, saturate it with water and leave it to drain, we cannot expect a quantity of water equivalent to the specific yield to drain from it. If the rock is coarse-grained, with correspondingly large pores, *some* water will drain from it, but not the whole of the specific yield. There will always be a layer at the lower side of the sample where, as in the capillary fringe, the pores are completely filled with water; in this layer, in simple terms, the weight of the water is balanced by the retention forces holding the water in the capillary-sized voids. Because of this effect, the determination of specific yield using small samples is difficult, and can require elaborate equipment.

To summarise, we have seen so far that we have a saturated zone, whose upper boundary is the water table; above this is the capillary fringe. Both the saturated zone and the capillary fringe are characterised by the fact that all their voids are water-filled.

Above the capillary fringe we have the **intermediate zone** in which the water content is generally at the specific retention value, in the form of thin films surrounding particles and lining the sides of pores; this water is called **pellicular water**. Any excess water in this zone drains under gravity towards the water table.

Above the intermediate zone, which may vary in thickness from zero to more than a hundred metres, is that complex fragment of the Earth's crust, the soil. The capillary fringe, the intermediate zone and the soil are all part of the unsaturated zone, in which surface tension effects play a major rôle. These effects, and the concepts of specific retention and specific yield, are all relevant to the soil zone, but the relationship between mineral particles and water in the soil is made more complex than elsewhere by the activities of plants. For this reason, and because the soil is the first control on infiltration as it starts its downward movement, the soil deserves a chapter to itself.

Selected references

Freeze, R. A. and J. A. Cherry 1979. *Groundwater*. Englewood Cliffs, N J: Prentice-Hall. (See especially Ch. 2.)

Holmes, A. 1965. *Principles of physical geology*, 2nd edn. Walton-on-Thames: Nelson.

Rodda, J. C., R. A. Downing and F. M. Law 1976. *Systematic hydrology*. London: Newnes-Butterworths. (See especially Ch. 4.)

Tabor, D. 1969. *Gases, liquids and solids*. Penguin Library of Physical Sciences. London: Penguin. (Contains – pp. 212–19 – an excellent presentation of surface-tension and capillarity theory.)

5 Soil water

To say that the soil deserves a chapter to itself is an understatement. The soil accounts for only the top metre or so of the Earth's crust, but it has probably attracted as much attention as the rest of the Earth's crust put together. Soil physicists and chemists work alongside biologists and agriculturists, and the products of their labours would fill a library, let alone a chapter. The reasons for all this interest are not difficult to find. The soil is our plane of contact with our planet; we live on it, and when we die we become part of it. Plants grow on and in it, and land plants directly or indirectly provide most of our food and oxygen.

The soil is a complicated mixture of organic and inorganic material, a curious blend of life and decay. It consists essentially of mineral particles, dead organic matter, living organisms, water and air. The mineral particles are derived from the decomposition of rocks by the processes known collectively as **weathering**; the rock debris may stay more or less in place or be transported by ice, wind or water. In either case the mantle of rock waste provides a base for living organisms. Bacteria are probably the first to arrive, followed by mosses and lichens. As these die and begin to decay, their remains form the first humus. Seeds are brought by the wind or by animal activity, and a cover of vegetation begins to form. The roots of plants help to bind the mineral particles together, and decaying plant material provides more humus. Earthworms and burrowing animals help to distribute the organic materials through the soil, and provide passageways for water and air. The respiratory activity of micro-organisms increases the carbon-dioxide content of the soil atmosphere, and provides a large proportion of the carbon dioxide which enables rain water to dissolve calcium carbonate.

On steep slopes the rock waste tends to be carried away before plants can establish a foothold to grow and bind it together, which is why such slopes are frequently bare rock or scree. On shallower gradients the type of soil and its rate of development will depend largely on the climate and on the type of rock from which the minerals are derived.

Limestones, as we saw earlier, are dissolved by rain water. Weathering of limestones therefore leaves little in the way of waste; only a few insoluble impurities remain, so that on limestone areas characteristically thin soils are formed. Sandstones weather to sandy soils which are usually permeable and which drain easily. Clays and shales generally

Soil water

(a)   (b)

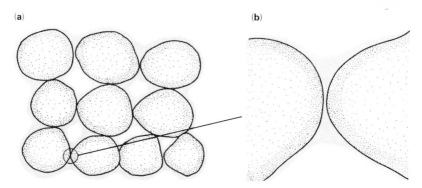

Figure 5.1 Specific retention   Films of water are held around grains by surface tension and by molecular forces.

weather to give heavy soils of low permeability, which may require careful tilling and extensive artificial drainage to make them productive.

The characteristic feature that turns rock waste into soil is the presence of living organisms, and these need water. Water generally enters the soil by infiltration, and occurs as films around the solid particles, held by molecular and surface tension forces (Fig. 5.1); air is also present in the soil voids. The thickness of the water films, and therefore the proportion of water to air, depends on the **degree of saturation** of the soil-pore space or on what is sometimes called the prevailing soil-moisture condition.

Suppose that we start with a soil that has just experienced heavy rainfall, so that all the voids are completely filled with water; the soil is saturated. We then allow the soil to drain under gravity (a process that is generally more or less complete in from two to five days, although it is doubtful whether it is ever truly complete). The water that has drained out by the end of this time represents the specific yield of the soil, and the remaining water the specific retention; under this condition the soil is said to be at **field capacity**, which is another way of saying that it is holding all the water which it can hold against gravity.

In reality, of course, drainage to field capacity does not occur in this ideal way, because water is evaporating from the soil surface and plant roots are extracting water from the soil more or less continuously. Once gravity drainage has virtually ceased, the plant roots begin to draw on the specific retention, the moisture content of the soil becomes reduced below the field-capacity figure, and a **soil-moisture deficit** begins to develop. This soil-moisture deficit is usually expressed in terms of the millimetres of rainfall that would have to infiltrate the soil to restore it to field capacity.

If we look closely (Fig. 5.1b) at the film of moisture around the soil particles in Figure 5.1a, we see that the interface between air and water is generally curved, with a concave surface presented towards the air; it is in fact a form of meniscus. We saw in Chapter 4 that under these conditions there will be a difference in pressure between the two sides of the interface, with the lower pressure on the convex side. In the soil the air on the concave side is in contact with the atmosphere and must therefore be at atmospheric pressure; it follows that the water on the convex side is at a pressure below atmospheric. It is this reduced pressure, or suction, that holds the water around the grains and that prevents it from draining under gravity, in just the same way as the water in the capillary tube in Figure 4.2a is prevented from draining.

When the soil is completely saturated there is no air in the voids and therefore no air/water interface; no suction exists. As soon as drainage begins and air enters the voids, menisci form and the water is under reduced pressure. Gravity drainage continues until the pull of gravity on the water films is insufficient to overcome the surface-tension forces, or suctions, holding the films around the soil particles; this is the field-capacity condition, the dividing line between specific yield and specific retention − a line which, in practice, is difficult to draw.

The suction of the water films is referred to by various names such as 'soil-water suction', 'pore-water suction' and 'soil-moisture tension'; the last term is especially popular in books on soil physics.

As the length of time since the last rainfall increases, more and more of the soil water − the specific retention − is used up, either directly as evaporation from the soil surface or by plants. As this happens the films of water around the soil grains become thinner, the larger pores become drained and the air/water interfaces become more sharply curved (Fig. 5.2). This means that the pore-water suction increases, and makes it progressively more difficult for plants to continue extracting water, because there is effectively a limit to how hard the plants themselves can 'suck'. To begin with, the plants respond by using less water − the stomata (openings) in their leaves and stems close so that less water is transpired.

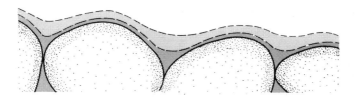

Figure 5.2 Pore-water suction   As water is withdrawn from soil the curvature of the menisci, and therefore the suction holding the water in place, increases as the films of water become thinner.

The reduced throughput of water also means that the plants' rate of growth is reduced.

Herbaceous plants, which have non-woody stems, stay rigid and upright because their cells are filled with water. If these plants are deprived of water for a long period then, rather as a car tyre goes soft and flat if the air is let out, so the cells begin to lose their rigidity and the plants begin to **wilt**. If water becomes available again within a reasonable time the cells can recover; if not, **permanent wilting point** is reached and the plants die.

The pore-water suction that exists at the time a plant wilts varies from plant to plant, but typically at the time permanent wilting point is reached the pore-water suction is theoretically capable of supporting a column of water over a hundred metres high. This shows how effective plants are at withdrawing water from the soil.

Under the influence of these high suctions the soil water may start to redistribute itself. We saw that downward drainage of the soil water ceases early in the drying process; as drying continues and suctions in the root zone increase, water may begin to move upwards to replace the water being removed by evapotranspiration. This upward movement of water will occur first from lower in the soil zone and then from the intermediate zone; in some cases, water may even be drawn up from the saturated zone.

When a commodity is scarce, governments sometimes ration it to ensure that everyone gets a fair share and to encourage people to use no more than they really need. When water becomes scarce, nature seems to impose her own system of rationing. To begin with, downward movement of water to the water table slows down as the soil drains to the field-capacity condition. Then, as more of the specific retention is used up, increasing pore-water suctions make it harder for plants to extract water; they are therefore obliged to reduce their consumption by transpiring less and slowing down their growth rate.

However, all this would be to little avail if the high pore-water suctions in the soil zone could simply cause water to move up from the water table; to counter this nature has one more trick up her sleeve. As suctions increase it is the larger pores that are drained first. Eventually water remains in the soil only as thin films held around particles by surface tension or by strong molecular forces. Only the smaller pores within the soil remain completely filled; the larger pores, which provide the easiest paths for water movement, are empty and unable to make any effective contribution to permeability. It therefore becomes progressively harder for water redistribution to occur and the upper part of the soil effectively becomes a barrier to further loss of water from the ground.

Pore-water suctions are obviously important to agriculturists and soil scientists. Several methods of directly or indirectly measuring these suctions are available; none of the methods is particularly easy or reliable. In general, the greater the suction to be measured, the greater the problem of measuring it. Many of the methods involve placing a water-saturated block of artificial porous material in contact with the soil. Water will be drawn from the block into the soil until the pore suction in the block is equal to that in the soil; the suction in the block is then measured with some form of vacuum gauge. At high suctions vacuum gauges cannot be used, and indirect methods are resorted to. One popular technique relies on the change in electrical resistance of a porous block as its pore-water suction, and therefore its water content, changes. The block is calibrated in the laboratory by noting the value of the resistance as various known suctions are directly or indirectly applied; measurement of the resistance of the block when placed in the soil enables the suction of the soil to be deduced.

The fact that evapotranspiration is reduced as the soil dries out is important to hydrologists, because evapotranspiration is one of the items in the water balance. There are various methods for estimating evapotranspiration; they make use of measurements such as temperature and wind speed, or of measurements of evaporation from water-filled containers of fixed size and shape. None of the methods is perfect; one will work better under one set of conditions, another under other conditions.

These methods calculate **potential evapotranspiration**, which is the amount of water that would evaporate, under the prevailing weather conditions, from short vegetation (such as mown grass) completely covering the ground and well supplied with water. Correction factors, which have been determined from field observations, are available to allow us to calculate the evapotranspiration from any other type of vegetation cover if we know the potential evapotranspiration value. Evapotranspiration from a forest of tall pine trees, or from a rice paddy field, for example, will be much more than that from short grass. The hydrologist must consider this when working out the water balance.

What happens when the vegetation is not well supplied with water? We have already seen that evapotranspiration falls below the potential value, as the plants begin to cut down on their use of water. A useful concept here is the **root constant** proposed by H. L. Penman; this is theoretically the value that the soil-moisture deficit can reach before the plants begin to use less water, i.e. before *actual* evapotranspiration falls below *potential* evapotranspiration. Because some plants are more effective than others at withdrawing water from the ground, the root constant differs from plant type to plant type; it can also vary with the type of soil. For most vegetation, it seems that until the soil-moisture deficit exceeds the root constant by about 25 mm, the rate of evapotranspiration falls only a little below the potential value; thereafter, it declines more rapidly, falling almost to zero as the wilting point is approached.

In practice, plants tend not to follow the rules that we make for them. The agriculturist working out soil-moisture deficits or the hydrologist working out a water balance may assume that a grass area is transpiring at its potential rate when the soil-moisture deficit is, say, 74 mm and at fifteen per cent less than its potential rate when the deficit increases to 76 mm; the grass, not being aware of their computer programs, is unlikely to oblige so precisely! We have to realise that concepts like soil-moisture deficit, potential evapotranspiration and root constant are *only* concepts – concepts that we have invented to help us to understand how nature operates, not the rules that nature actually follows.

Many soil scientists now have reservations about the use of concepts like these. Beginning with field capacity, they argue that there is no fixed value of water content or pore-water suction at which gravity drainage of water from pores will completely cease; hence we cannot accurately define specific yield or specific retention. Further, since field capacity is the standard with reference to which we define soil-moisture deficit, it looks as though that must go out of the window too, taking root constant with it. But the test of concepts like these is whether they enable us to make useful and reliable predictions. So long as they help rather than hinder our understanding – which in practice probably means so long as we realise their shortcomings – then these concepts are probably worth retaining, at least until there is something better to put in their place.

Vital as the soil and its water may be to the farmer, what really matters to the hydrogeologist is how much water manages to get through it to recharge the aquifers. Soil is important to us because it is effectively the first control that water meets when it starts to enter the ground. To see this, let us look in more detail at this stage of the water cycle, which was considered briefly in Chapter 3.

Suppose a heavy fall of rain occurs after a long period of dry weather. As rain begins to fall on the dry surface of the soil, the pore-water suctions assist gravity in drawing the water down into the soil, so that it is absorbed readily; the infiltration capacity is very high. As rainfall continues the infiltration capacity will almost certainly fall. The reasons for this include the compacting effect which the raindrops themselves may have on the soil surface, tending to close the pores, and of course the fact that the pore-water suctions will decrease as the pores begin to fill with water. Water will still be absorbed by the soil, however, so long as the infiltration capacity exceeds the rate at which the rain is falling — the **rainfall intensity.**

It may occasionally happen that the rainfall intensity exceeds the infiltration capacity, resulting in the Hortonian overland flow discussed in Chapter 3. It is possible for this to occur even when a high soil-moisture deficit is present — even though the soil needs water, it cannot absorb it fast enough to prevent some water running away over the ground surface. On thick soils, with a cover of vegetation, this process is rarely observed — although it must be said that the heavier the rain, and therefore the more likely overland flow, the less likely there is to be anyone there to observe it! The presence of gullies (water-worn channels) on steep slopes in tropical and sub-tropical areas is evidence that overland flow occurs there; the general absence of such gullies in north-west Europe suggests that it is unusual here. In areas like Britain, the controlling factor on infiltration and the cause of overland flow is likely to occur in the soil profile rather than at the surface: this is the rate at which water can move down through successive layers of the soil to leave space for more water to enter the layers above. The effect is most important near stream channels, where the combination of interflow — becoming concentrated as it approaches the channel — and of a shallow water table can result in complete saturation of the soil profile; any further water arriving at the ground surface is thus compelled to travel over the soil as overland flow.

It might seem impossible at first sight that recharge to an aquifer should occur when a soil-moisture deficit exists. It would appear impossible for water to pass downward through soil which has a high pore-water suction until the suctions have been satisfied and the soil restored to field capacity. However, there seems little doubt now that aquifers do receive recharge from rainfall even when soil-moisture deficits are present. The most likely explanation seems to be that many soils crack on drying out. The cracks provide a path for rain water to enter the deeper layers of the soil or intermediate zone, bypassing the shallower layers where a soil-moisture deficit may exist.

Rainfall that is not heavy but gentle and prolonged provides ideal conditions for recharge to occur. The rainfall intensity is low enough not to exceed the infiltration capacity nor to cause partial saturation and interflow; then, provided that the underlying materials are permeable, once the soil-moisture deficit is satisfied (or even sooner if cracks are present), water can move down to recharge the aquifer. What it finds when it gets there — well, that's the subject of the next chapter.

Selected references

Childs, E. C. 1969. *An introduction to the physical basis of soil-water phenomena.* London: Wiley.

Freeze, R. A. and J. A. Cherry 1979. *Groundwater.* Englewood Cliffs, NJ: Prentice-Hall. (See especially pp. 39–44 and Ch. 6.)

Holmes, A. 1965. *Principles of physical geology,* 2nd edn. Walton-on-Thames: Nelson.

Rodda, J. C., R. A. Downing and F. M. Law 1976. *Systematic hydrology.* London: Newnes-Butterworths. (See especially Ch. 4.)

Ward, R. C. 1975. *Principles of hydrology,* 2nd edn. Maidenhead: McGraw-Hill.

6   Groundwater in motion

So far in this account I have talked about water entering aquifers, moving through them and leaving them, without really discussing how this process occurs. Now it is time to consider groundwater movement, and its causes, in more detail.

Although this chapter is far from a rigorous treatment of groundwater flow, it does go into more detail than most of the others in this book. I make no apology for this, as it seems to me that an understanding of these basic physical principles (even in elementary terms) is essential to an understanding of hydrogeology. I have summarised the most important points in simplified form at the end; if you find the chapter heavy going, read the Summary and perhaps return to the more detailed sections afterwards.

Water movement

To understand how groundwater moves we first need to think about the movement of water in general − indeed, we need to think about what makes anything move at all. If you feel like getting up and running about, you might say that you are feeling energetic; if on the other hand you felt particularly listless you could say that you were lacking energy. Energy is simply the capacity to do work. A crude way of measuring how energetic you feel is to see how many flights of stairs you can climb − in other words, how high you can raise yourself above a given level. Climbing stairs is a form of work.

Water, like people, needs energy to make it move; one way of expressing the energy of water, like that of people, is to measure the height to which it can raise itself above an arbitrary given level or **datum level**. We call this height the **hydraulic head** or simply the **head** above that datum. The datum can be arbitrary because, as we shall see, we are mainly interested in comparing the energy of water in one location with that in another. Provided that we measure the energies relative to the *same* datum, the position of the datum itself is irrelevant.

Consider the arrangement in Figure 6.1. Tube B contains more water than tube A, but experience tells us that if we open valve C water will flow from A into B, until the water level in both tubes is at the same height above the datum level. We know that flow occurs from A to B

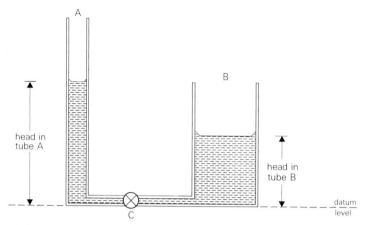

Figure 6.1   Head differences   There is a head difference between the water in tubes A and B. The *head* of water in A is greater than that in B, even though B contains a greater *volume* of water. If valve C is opened water will flow from A to B.

because initially the water level in A is higher than in B. In everyday terms we speak of 'water finding its own level'. More scientifically we can say that there is a **head difference** between A and B, and that the head

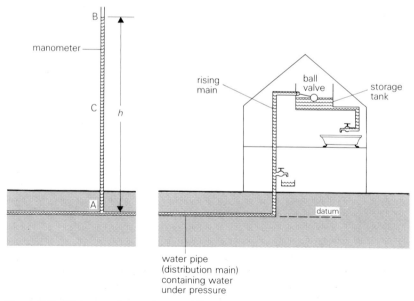

Figure 6.2   Water supply to a house   Water will flow from the underground supply pipe up the rising main to the tap and storage tank because it has pressure energy. The height *h* to which it would rise in a manometer inserted at A is a measure of its 'pressure head'.

at A is greater than that at B. Flow occurs to equalise these heads. Is head, then, simply related to height above the ground? In other words, does water always flow downhill? Unfortunately, no – or perhaps I should say fortunately, no, because if water could not flow uphill we should have great difficulty in distributing it around the countryside to meet domestic, industrial and agricultural demand.

Consider the simple case of the water supply to a house (Fig. 6.2). The water supply pipe is underground, yet the water travels up the rising main to the taps or into the storage tank, which is usually in the roof cavity. Clearly this could not occur if head were simply a function of height. The upward flow takes place, of course, because the water in the pipe is under pressure. If we insert a **manometer** (an open-topped pipe, of large enough diameter for capillary effects to be negligible) into the water main at point A, water will rise up the pipe to point B to give us a measurement of the **pressure head** at A. So long as B is above the level of the ball valve, water can flow up into the storage tank.

We therefore see from Figure 6.2 that a head difference can be a result of *pressure* difference, as well as a result of a difference in elevation. To distinguish between them, it is quite in order for us to speak of 'elevation head' and 'pressure head'. But are these two types of head really different? To answer this question, we need to go back a little, to the idea of head as a measure of energy.

We define hydraulic head of water as the energy of the water per unit weight. Water can possess energy in several ways; one of these ways is **elevation energy**, which is energy possessed by virtue of position (Fig. 6.3). A mass m a vertical distance h above a datum has elevation energy

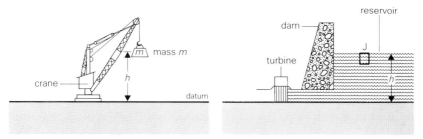

Figure 6.3 Elevation energy A mass m at a height h above a datum possesses elevation energy of mgh units relative to that datum. This is the work that a crane would have to do to lift the mass from the datum (in this case, ground level) to that height: it is also the amount of work that the weight could be made to do in returning from height h to the datum. The water in a reservoir also possesses, energy. The amount of work that each unit volume of water can be made to do in passing through the turbine is proportional to h. A unit volume of water has mass ϱ, so the unit volume at J has elevation energy of ϱgh. Its weight is ϱg, so its *head* (energy/unit weight) is simply h.

mgh units relative to that datum – *mgh* units is the amount of work needed to lift the mass from the datum to its position, or the amount of work which we could make it do in allowing it to move from that position to the datum. (As usual, *g* is the acceleration due to gravity.) A unit volume of water has a mass of ϱ, where ϱ is the density of water; a unit volume of water at height *h* therefore has elevation energy ϱ*gh* units. Since the weight of a unit volume of water is ϱ*g*, it follows from our definition of hydraulic head (energy per unit weight) that water at height *h* above the datum is at an elevation head of ϱ*gh*/ϱ*g*, or *h*. All of which may seem a complicated way of stating the obvious, but it is sometimes advisable to make sure that what is 'obvious' is really based on fact!

Let us now think about pressure. We saw from Figure 6.2 that when we insert a manometer tube into our water main at point A, the water rises up the tube. Clearly then, water under pressure possesses energy and therefore hydraulic head. What we have to do is express this in quantitative terms.

In Figure 6.2, when we have connected our manometer and when the water is no longer flowing, we have a situation in which the water that has moved from A to B has exchanged the pressure energy it possessed at A for the elevation energy it possesses at B. We know from the law of the conservation of energy that these two quantities of energy must be equal; since the water at B has energy ϱ*gh* per unit volume and therefore has a head of *h* units, then the water at A must have the same energy and therefore the same head, *h*. If this were not so, flow would take place in the manometer.

In general, the pressure at any depth *z* below the surface of a liquid is higher by an amount *p* than the atmospheric pressure acting on the liquid surface, where

$$p = \varrho g z.$$

For convenience, we can adopt a convention of measuring all pressures relative to atmospheric pressure. Thus in Figure 6.2 the pressure *p* at A is equal to ϱ*gh*, because A is at depth *h* below the water surface. Since the head at A equals the head at B, which in turn equals *h*, we see that the head at A is given by

$$h = \frac{p}{\varrho g}.$$

This is a general formula for the calculation of pressure head.

Pressure energy and elevation energy are both forms of **potential**

energy, which in general terms is energy which a body possesses by virtue of its position or state. Water can possess both these forms of energy simultaneously; the water at point C in the manometer tube in Figure 6.2, for example, possesses some elevation energy as a result of its height above the datum and some pressure energy as a result of its depth below the water surface.

Elevation energy is potential energy possessed by virtue of position. Pressure energy is potential energy possessed by virtue of state; it is somewhat analogous to the energy of a compressed spring. A clockwork toy, set to run slowly up a slope, converts the potential energy of its spring to elevation energy. Similarly, in Figure 6.2, water flowing from the water main to the storage tank converts its pressure energy into elevation energy.

In any body of standing water, the pressure energy increases with depth below the water surface and the elevation energy increases with height; the sum of the two is constant – if it were not, the water would be in motion. These relationships for static water in a reservoir are shown in Figure 6.4. At any point K, the sum of pressure head h_p and elevation head h_e is equal to the static head h_s, which is the height of the water surface above the datum.

There is a third way in which water can possess energy – by virtue of

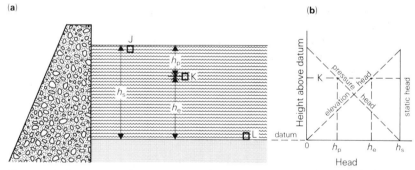

Figure 6.4 Pressure head and elevation head in standing water, such as a reservoir
(a) A unit volume of water at the water surface (e.g. volume J) possesses elevation head but no pressure head; at the datum level (which in this example is at the bottom of the reservoir) a unit volume of water possesses pressure head but no elevation head (e.g. volume L). At any intermediate depth the water (e.g. K) possesses some pressure head h_p and some elevation head h_e. The sum of the two components is the static head, h_s, and is equal to the height of the water surface above the datum. (b) A graphical representation of the head conditions in (a). The elevation head decreases with depth, whereas the pressure head increases with depth; at any depth the sum of the two is equal to the sum at any other depth and to the static head, h_s. Because pressure increases with depth the dam is made thicker towards its base.

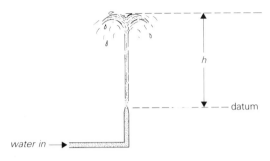

Figure 6.5 Energy changes in a vertical jet of water On leaving the nozzle, water possesses no elevation energy relative to the datum, but does possess some kinetic energy. At the top of the jet the water's velocity (and therefore its kinetic energy) is momentarily zero but its kinetic energy has been converted into elevation energy. Pressure energy (the water is at atmospheric pressure from the time it leaves the nozzle) is unchanged throughout.

movement. Moving water clearly possesses energy – a horizontal jet of water, for example, can do work by rotating a turbine wheel. This energy is called **kinetic energy** and the contribution which it makes to the total hydraulic head is called the **velocity head** or **dynamic head**.

A mass m moving with speed v has kinetic energy of $\frac{1}{2}mv^2$. From our definition of hydraulic head, the velocity-head contribution of unit volume of water (mass ϱ) is therefore $\frac{1}{2}\varrho v^2/\varrho g$ or $v^2/2g$. We can demonstrate this using, once again, the law of the conservation of energy.

Suppose we direct a jet of water vertically upwards (Fig. 6.5). The water leaves the jet with speed v, and we draw our energy datum line level with the nozzle. Once the water leaves the nozzle it is at atmospheric pressure, so there are no pressure-energy changes. Where it leaves the nozzle it is level with the elevation-energy datum, so it possesses only kinetic energy, $\frac{1}{2}\varrho v^2$ per unit volume. At the top of the jet, the water is effectively stationary, so its only energy is elevation energy, ϱgh per unit volume. Assuming that no energy is used in overcoming friction, for example with the air, then these two quantities must be equal, i.e.:

$$\tfrac{1}{2}\varrho v^2 = \varrho gh$$

whence

$$h = \frac{v^2}{2g},$$

which is what we wanted to demonstrate.

Suppose now that we set up a horizontal long straight length of pipe,

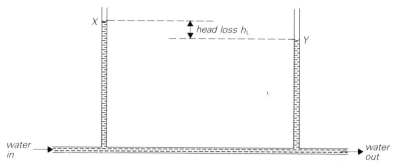

Figure 6.6   Steady flow through a horizontal pipe   The head loss between X and Y represents energy that has been converted into heat.

whose cross-section is the same throughout its length (Fig. 6.6). Near each end we fit a manometer. We then pass a steady flow of water through the pipe — by steady, we mean that in any instant the same volume of water enters the pipe as in any other instant, and the volume leaving the pipe in any instant is the same as the volume entering it.

If we carefully observe the water levels in the manometers, we shall see that the water level in the manometer nearest the outflow end of the pipe is lower than the level in the manometer at the inflow end. Surely, at first sight at least, this is a paradox? The pipe is horizontal, so there can be no change in elevation energy as the water flows along it; the flow is steady and the cross-section is uniform, so the speed and therefore the kinetic energy must also be constant. Yet there is a reduction in head, which can therefore only be the result of a decrease in pressure. The decrease in pressure energy has not been balanced by an increase in kinetic or elevation energy, so where has this energy gone? We know from the law of energy conservation that it cannot have been destroyed.

The answer is that the energy has been converted into heat, as a result of friction between the water molecules themselves and — to a lesser extent — between the water and the surface of the pipe. (In basic terms heat is really a form of kinetic energy — when a substance is heated its molecules vibrate more rapidly. It is therefore easy to see how friction — which can be thought of as an interaction between molecules — increases molecular kinetic energy and therefore generates heat.) In just the same way that you become hot if you run around for a time, or that a bearing in a motor becomes warm with use, so the water and pipe will have become heated, although usually by so small an amount that we should not be able to measure the resulting rise in temperature. The sad thing about the energy that has been converted into heat in this way is that it is irrecoverable. We cannot, for example, cool the pipe down and increase the pressure of the water again. Therefore although strictly speaking no energy has been destroyed, *useful* energy has been 'lost' from the system, and for this reason the fall in head between the two manometers in Figure 6.6 is commonly referred to as **head loss**.

The faster you run, the more heat you generate within your muscles; the faster we move water through a pipe, the more energy is dissipated as heat and the greater the head loss along the pipe. Head has units of length, and we can divide the head loss between the ends of the pipe by the length of the pipe to obtain a quantity called the **hydraulic gradient**, or **head gradient**. Other things being equal, the faster we want the water to move, the steeper this gradient must be. In one way water always does flow downhill, but the 'hill' it flows down is the hydraulic gradient – the steeper this gradient, the faster the water will flow. If, in Figure 6.6, we stop the flow altogether the gradient becomes zero – in other words the heads in the manometers become equal; the water 'has found its own level' again.

If we were to close off the outflow end of the pipe in Figure 6.6 the water levels in the manometers would rise; whether or not they actually overflowed would depend on whether or not they were higher than the head at the inflow point. Let us assume for a moment that the manometers *are* higher, so that now we have a condition of no flow through the pipe, with the heads in the manometers equal. Why are these heads now higher than they were when water was flowing through the pipe? To answer this, we look again at the energy conditions.

When water was flowing through the pipe, in addition to some energy being dissipated as head loss, the flowing water possessed kinetic energy. The manometers do not measure total head, but **static head**, which in turn is proportional to potential energy. When the water begins to move, some potential energy is converted into kinetic energy; the *total* head (except for that part dissipated as heat) remains unchanged, but the *static* head decreases.

We see this very clearly if water flows steadily through a horizontal pipe containing a constriction (Fig. 6.7). For water to flow through this constriction at the same rate as through the rest of the pipe, it has to speed up, so that the volume flowing through the constriction in unit time is the same as the volume flowing through any other part of the pipe in unit time. This increase in speed means an increase in kinetic energy, which in turn must mean a reduction in potential energy. We measure this as a reduction in static head between manometer A and manometer B, but in this case only part of the head loss is due to friction, the remainder being caused by the change from potential to kinetic energy.

The elevation energy of the water is constant, so the reduction in potential energy must result from a decrease in pressure. It may seem

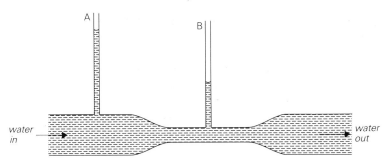

Figure 6.7 Pressure decrease at a constriction When water flows through a constriction its velocity (and therefore its kinetic energy) increases. If, as in this case, the pipe is horizontal, then the elevation energy is unchanged; therefore pressure must decrease to compensate for the increase in kinetic energy. This pressure reduction is seen as a head difference between manometers A and B.

remarkable that pressure should decrease inside a constriction, but it is a fact. The clockwork-toy analogy may be useful again here. If we set our toy running quickly along a horizontal surface, the spring runs down as its energy is converted into kinetic energy and used to overcome friction. If we wished the toy to go faster, the spring would have to wind down more rapidly to supply the extra kinetic energy; the energy in the spring is analogous to the pressure in the pipe. This pressure change in water flowing through a constriction provides a means of measuring the rate of flow of water, using an instrument called a Venturi meter.

On first reading, all this may seem rather complex. The situation looks worse when we stop to think that so far we have been talking mainly about water flowing through a smooth, circular pipe; how much worse things must be when we start to deal with water flowing through the complex twists and turns that make up the pore space in, say, a sandstone! Paradoxically, things are both worse and better. If we tried to describe the water movement along all the individual flow paths between the sand grains, then we should find the task beyond us. Fortunately, we have no need to do this; what interests us is the total flow resulting from water movement through the combination of all these flow paths: this is much simpler. Another simplifying fact is that groundwater generally moves very slowly – so slowly that its kinetic energy (and therefore its velocity head) is usually negligible.

Darcy's law

In studying the flow of water or of any other fluid through rock, we make use of a law, called Darcy's law, formulated more than a hundred years ago by a Frenchman. Henri Darcy was born in Dijon in 1803 and trained as an engineer. Dijon had long suffered from the lack of a dependable supply of safe drinking water, and shortly after graduating Darcy began work on an attempt to solve the problem. He designed a collection system for water from a large spring more than 10 km from Dijon, piped the water to the city and arranged for its distribution to standpipes. For the first time in its history, Dijon had a reliable water supply.

This practical success alone was a considerable achievement, but Darcy went further. In Paris he carried out a great deal of scientific work on the problems of water distribution, including studies of pipeflow and the measurement of rate of flow. An account of his work at Dijon, together with many of his experimental findings, was published in 1856 under the title *Les fontaines publiques de la ville de Dijon* – the public water supply of Dijon. One chapter dealt with the purification of water by filtering it through beds of sand. It was during the course of this work that Darcy derived the relationship that bears his name.

If we wished to test the relationship for ourselves, we could use an arrangement like that shown in Figure 6.8. Here we have a cylinder of

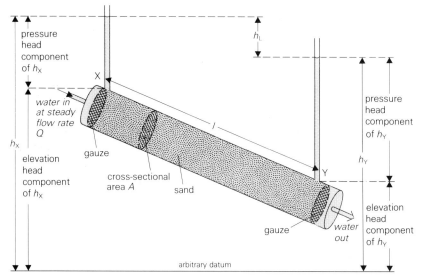

Figure 6.8 Experimental verification of Darcy's law The flow rate Q is proportional to the cross-sectional area A and the head loss h_L, and inversely proportional to the length l, i.e. it is proportional to A and to the hydraulic gradient h_L/l. The head loss h_L is the difference in head between the two manometers X and Y, which are a distance l apart (l is measured along the direction of flow). The head h_X at X consists of an elevation-head component and a pressure-head component; the same applies to h_Y at Y. Notice that in this example the water is flowing from a region of lower pressure to one of higher pressure.

cross-sectional area A, filled with sand which is held between two mesh screens. Water flows through the cylinder at a steady rate Q, measured for example in litres per day. Manometers at X and Y, a distance l apart along the cylinder, measure the head of the water relative to the datum level; the difference between the two heads is the head loss, h_L.

If we double the flow rate Q we find that h_L doubles also – we have to use twice as much energy to drive the water through the sand twice as fast. If we halve Q, we halve h_L; in general we find that the flow rate is directly proportional to the head loss, i.e.

$$Q \propto h_L.$$

Suppose we now vary the length of our sand column, but keep the same head loss between the ends. We find that as l increases, Q decreases, and vice versa, in strict proportion. If we double l, we halve Q. In general,

$$Q \propto \frac{1}{l}.$$

Combining these two results, we see that

$$Q \propto \frac{h_L}{l}.$$

We met the term h_L/l when discussing Figure 6.6 and called it the hydraulic gradient. We see therefore that the rate at which water flows through sand is proportional to the hydraulic gradient.

If we put another identical cylinder beside the first, filled with identical sand and with flow occurring in the same way, we should have in effect twice the flow occurring through twice the cross-sectional area. It seems reasonable to assume, as Darcy did, without actually performing the experiment that

$$Q \propto A.$$

In summary therefore we have

$$Q \propto A \frac{h_L}{l}$$

which is the same as saying that

$$Q = KA \frac{h_L}{l} \tag{6.1}$$

where K is a constant of proportionality.

Since we can measure Q, A, h_L and l in our experiment in Figure 6.8, we can calculate the value of K. If we did this, and then emptied all the sand out of the apparatus, replaced it with finer sand, repeated the experiment and then calculated K again, we should almost certainly find (as Darcy did) that the new value of K was smaller than the first one. If we used coarser sand, we should find that K was larger. K is the permeability, or more strictly the **hydraulic conductivity**, of the sand; coarse sand is more permeable than fine sand because the spaces between coarse grains are correspondingly larger and allow an easier passage for the water.

Equation 6.1 is **Darcy's law**, which, expressed in simple terms, says that a fluid will flow through a porous medium at a rate which is proportional to the product of the cross-sectional area through which flow can occur, the hydraulic gradient and the hydraulic conductivity.

If we repeated our experiments with the sand-filled cylinder but used syrup, say, we should expect, other things being equal, to find the flow rate much less than when we used water. This would mean, since everything else was unchanged, that the value of K that we calculated using

---

Darcy's law and other laws of physics

You may recognise the similarity between Darcy's law, Fourier's law of heat transfer, and Ohm's law. Ohm's law is usually expressed as

$$I = \frac{V}{R}$$

where I is electric current, R is resistance and V is the potential difference (the 'voltage') across the resistor. If we consider a cylindrical wire, length l, cross-sectional area A and conductivity c, then

$$R = \frac{l}{cA}$$

so that

$$I = cA\frac{V}{l},$$

and the similarity with Darcy's law is obvious. Clearly current is analogous to rate of flow of water, and potential difference V is analogous to head loss h_L. This analogy is used in resistance-network models of aquifers, which will be discussed in Chapter 10.

Darcy's law would be lower. Clearly, then, the hydraulic conductivity K depends not only on the porous medium but on the fluid which is filling the pores and passing through them – a fact which Darcy probably understood, as he described K as '*dependant* on the permeability of the [sand] bed'.

The property of the fluid that affects hydraulic conductivity is the kinematic viscosity, which is usually denoted by the Greek letter ν (pronounced 'new'); as the kinematic viscosity of the fluid in a porous medium increases, the hydraulic conductivity decreases. The kinematic viscosity of water decreases as its temperature increases.

Most rocks are not uniform, or **homogeneous**. As a result of natural variations during their formation and subsequent alteration, they are **heterogeneous**: their properties vary from place to place within the rock. Figure 6.9a shows an example of vertical heterogeneity. At any point in the rock, the permeability will usually also vary with direction; in Figure 6.9b, for example, it is easier for water to flow between the grains in the *horizontal* direction (parallel to the bedding) than in the vertical direction. Equality of properties in all directions at a point is **isotropy**; variation of properties with direction is **anisotropy**. Most sedimentary rocks are anisotropic with respect to permeability because they contain grains which are not spherical but which are elongated in one direction or shortened in another. When deposited, these grains tend to settle with their shortest axes more or less vertical, and this can cause the permeability parallel to the bedding to be greater than that perpendicular to the bedding, by a factor that typically varies from one to four. When heterogeneity like that shown in Figure 6.9a is also taken into account, the bulk permeability (i.e. the permeability of a large block) of sediments

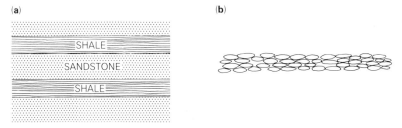

Figure 6.9 Heterogeneity and anisotropy (a) An example of vertical heterogeneity. It is easier for water to flow between the shale layers than across them. Therefore, although the sandstone and shale layers may individually be isotropic, the horizontal permeability of the combination of beds is greater than the vertical permeability. (b) Within a single rock type, platy grains tend to be deposited with their longer axes horizontal. This texture causes permeability anisotropy although the rock may be homogeneous.

Hydraulic conductivity and intrinsic permeability

The concept of hydraulic conductivity is a useful one in most topics of hydrogeology, since the viscosity of groundwater (which depends mainly on its temperature and the material dissolved in it) does not vary greatly from one aquifer to another. When considering water at great depths — where it is usually hot and saline — or rocks which contain fluids other than water, it is useful to be able to separate the properties of the rock from the properties of the fluid. This is particularly true when considering gas- or oil-bearing strata, since the viscosities of these fluids are very variable. For this reason the concept of **intrinsic permeability** was introduced (symbol k). The relation between hydraulic conductivity (or field permeability) and intrinsic permeability is

$$k = \frac{K\nu}{g} = \frac{K\mu}{\varrho g}$$

where ν is fluid kinematic viscosity, μ is dynamic viscosity, ϱ is density and g is gravitational acceleration. k thus has the dimensions of L^2 or area. A variety of units such as the cm^2 or μm^2 is possible, but the oil industry uses a unit called the **darcy** (plural *darcys*), defined as the permeability that permits a flow of 1 ml of fluid of 1 centipoise viscosity completely filling the pores of the medium to flow in 1 s through a cross-sectional area of 1 cm^2 under a gradient of 1 atm/cm of flow path.

Despite the rather appalling combination of inconsistent units, and the use of a pressure-gradient term instead of the hydraulic gradient that should strictly have been employed, the darcy and its derivative, the millidarcy (1 darcy = 1000 millidarcys), have become the almost universal units of permeability in the oil industry.

When we use the term 'permeability' we should, strictly, make it clear whether we mean k or K. But in practice, in hydrogeology, the variation in viscosity is usually so small as to make the distinction relatively unimportant. When we speak of one rock as being more permeable than another, it is tacitly assumed that they contain groundwater with similar properties and not that one is filled with water and the other, say, with syrup.

is usually much greater parallel to the bedding than perpendicular to it.

In order to talk about relative values of hydraulic conductivity, we clearly need to define it and give it some units. If we consider a homogeneous and isotropic porous medium with all the pores filled with a single fluid, the hydraulic conductivity is the volume of the fluid at the prevailing kinematic viscosity that will move in unit time under a unit hydraulic gradient through a unit cross-sectional area perpendicular to the flow. If we use Figure 6.8 as an example, unit hydraulic gradient means that h_L and l must be equal; then, if we make A equal to 1 m^2, the volume rate of flow through the cylinder is equal to the hydraulic conductivity (from Eqn. 6.1). Hydraulic conductivity thus has units of volume per time per area; hydrogeologists usually express it in terms of cubic metres per day per square metre, which reduces to metres per day (m/day). Civil engineers frequently use the centimetre/second (cm/s); the metre/second (m/s) is sometimes used, but is too large for most purposes.

The permeabilities of natural materials vary by many orders of magnitude. (An order of magnitude is a factor of ten.) Rocks that have hydraulic conductivities of 1 m/day or more when filled with ordinary groundwater are generally regarded as permeable and likely to form good aquifers; those with hydraulic conductivities of less than 10^{-3} m/day would generally be regarded as impermeable. However, all things are relative. A layer of sand with a hydraulic conductivity of 0.1 m/day might be regarded as impermeable if it separated two gravel strata with hydraulic conductivities of 50 m/day, but a similar sand occurring within clay with a hydraulic conductivity of 10^{-4} m/day might be regarded as permeable. A hydraulic conductivity of 10^{-3} m/day might be embarrassingly high to a civil engineer if it occurred in bedrock underneath a dam, because the high head of water behind the dam might cause significant flow through the rock.

It is worth emphasising at this point that in using Darcy's law we are, as was said earlier, forgetting about the tortuosity of the individual pore channels. We are instead assuming that we can carry out a form of averaging, replacing the twisting microscopic channels of the **porous medium** formed by our rock fabric with a macroscopic **continuum**, through which the flow will be the same as the resultant of all the flows through the microscopic channels of the rock. This may seem, intuitively, a logical thing to do – after all, we are in general interested in the bulk movement of water through the rock, not the minute flow through an individual pore channel – but to demonstrate the validity of the assumption in rigorous physical terms is far from straightforward. We can demonstrate Darcy's law in the laboratory, but to derive it mathematically from the fundamental equations of fluid mechanics is another matter. M. King Hubbert, an American geologist who did a great deal of pioneering work on the movement of groundwater and the migration of petroleum, appears to have been the first person to attempt the derivation with any degree of success – in 1956, to mark the centenary of Darcy's publication. Despite the attempts of Hubbert and others, Darcy's law remains essentially an empirical law, its validity resting on experimental evidence.

Applications of Darcy's law

How is Darcy's law used? In many ways. One of the most frequent and easiest to understand is in calculating the natural flow through an aquifer. Imagine a strip of aquifer (Fig. 6.10a), with flow occurring in the direction from borehole X to borehole Y. The hydraulic conductivity is K m/day. How much water flows through in a day?

The cross-sectional area A is simply bw. If the flow is so slow that the kinetic energy of the water is negligible, the hydraulic gradient can be determined from the difference in water levels in the two boreholes, which are effectively manometers. This difference is H, so the hydraulic gradient is H/l, and we see that this is simply the slope of the potentiometric surface.

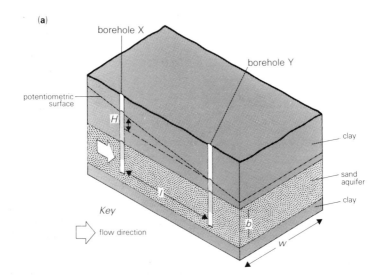

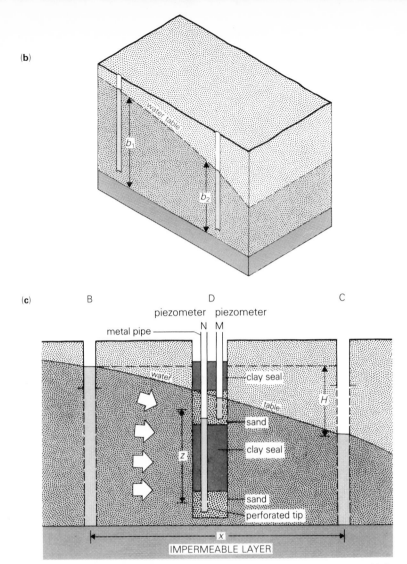

Figure 6.10 Flow through aquifers (a) In a *confined* aquifer the saturated thickness *b* is constant. The flow rate *Q* through the aquifer is (from Darcy's law) equal to *KbwH/l*. (b) In an *unconfined* aquifer, the sloping water table causes a change in the saturated thickness and hence in the area through which flow occurs. (c) In an unconfined aquifer the water table must be sloping if flow is taking place. Because of this slope (shown exaggerated here) there is a vertical component of flow in the upper part of the saturated aquifer which means that a vertical hydraulic gradient must be present in addition to the horizontal hydraulic gradient. The vertical hydraulic gradient can be measured by comparing water levels in the piezometers at M and N, which are open only to a small thickness of the aquifer; this vertical gradient is *h/z*. The horizontal hydraulic gradient can be measured by comparing water levels in B and C and is *H/x*.

From Darcy's law, the flow Q is given by

$$Q = KAH/l = KbwH/l.$$

Obviously, the effectiveness of a rock stratum as an aquifer depends not only on its hydraulic conductivity but on its cross-sectional area perpendicular to the flow direction – i.e. on its thickness b and width w. In particular, in carrying water to a well it is the product of hydraulic conductivity K and thickness b that is important. This product Kb, which is frequently given the symbol T, is called **transmissivity**. For the case shown in Figure 6.10a, Darcy's law can therefore equally well be written as

$$Q = TwH/l.$$

Transmissivity has units of hydraulic conductivity multiplied by thickness; it is frequently expressed in $m^3/day/m$, which reduces to m^2/day. Important aquifers such as parts of the Chalk and the Permo-Triassic sandstones have transmissivities in excess of $1000\ m^2/day$.

Sometimes the hydraulic gradient H/l is written as I, so Darcy's law may also be expressed as $Q = KAI$ or $Q = TIw$.

In the case of an unconfined aquifer (Fig. 6.10b) the slope of the water table is a measure of the hydraulic gradient. Darcy's law still applies, but horizontal flow is occurring only through the saturated part of the aquifer. In this case, the transmissivity is the product of hydraulic conductivity K and b, where b is the *saturated* aquifer thickness. A complication is that because the water table is sloping, b is not constant; in cases like this, either an averaging procedure has to be used before Darcy's law can be applied, or Darcy's law has to be applied to successive 'slices' of the aquifer perpendicular to the flow direction.

Further, because the water table is sloping, the flow is not purely horizontal. As Figure 6.10c shows, there must be a vertical component of flow in the upper part of the saturated portion of the aquifer; this in turn means that there is a vertical hydraulic gradient in addition to the horizontal one. If we observe the water level (which is a measure of the head) in boreholes at B and C we find that there is a difference H, which enables us to calculate the horizontal hydraulic gradient between B and C. If we construct boreholes M and N, at the same horizontal location D, but each open to the aquifer only at a particular and different depth, we find that there is also a difference in water level between them. Boreholes like M and N, which are sealed throughout most of their depth in such a way that they measure the head at a particular depth in the aquifer, are called **piezometers**. The head difference h divided by the vertical distance (in this case, z) between the two points at which the heads are measured, represents the **vertical hydraulic gradient** at D.

Because of these vertical hydraulic gradients, the level at which the water stands in a deep borehole may not be exactly the level of the water table. If the flow is downward, the head will decrease with depth, so the water level in a deep well will be below the water table. If the flow is upward, the head will increase with depth and the water level will stand above the water table. Layers of low permeability tend to exaggerate these vertical gradients; many so-called artesian wells – see Chapter 7 – are an extreme example of a condition of upward flow, with the water level in the well standing above ground surface.

In a homogeneous aquifer, the relationship $T = Kb$ is perfectly valid and straightforward. Most real aquifers, however, consist of a combination of layers of varying permeabilities and thicknesses: in this case, strictly, the transmissivity must be worked out by calculating the contribution of each layer – its individual permeability times its thickness – and adding all these contributions together. An 'average' permeability can then be determined by dividing the total transmissivity by the total thickness. The problem becomes particularly complicated in the case of fissure-flow aquifers like the Chalk, where the rock mass itself has negligible permeability and virtually all flow takes place through a few fissures. Here the idea of 'average' permeability is of limited value, since it will vary greatly according to the rock thickness over which it is averaged.

Flow to a well

One of the most important applications of Darcy's law is in considering flow to a well or borehole. Consider a borehole that goes into a confined, homogeneous, isotropic aquifer of hydraulic conductivity K and reaches to the bottom of the aquifer (Fig. 6.11a; such a borehole is said to be *fully penetrating*). Suppose that the potentiometric surface is initially horizontal so that the groundwater is not moving at all, in any direction – an unlikely state of affairs, but it will make the discussion easier! Then, using a pump inserted into the borehole, we start pumping water out of the borehole at a rate of Q m^3/day.

The action of the pump in withdrawing water from the well causes a reduction in pressure around its intake, and this in turn creates a head difference between the water in the borehole and that in the aquifer (Fig. 6.11b). Water flows from the aquifer to the borehole to replace that

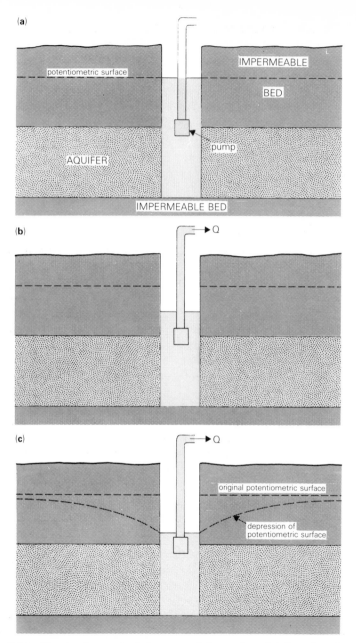

Figure 6.11 The development of the cone of depression (a) If no flow is taking place, the potentiometric surface is initially horizontal. (b) If the pump is started the head in the well is lowered. (c) This causes water to flow from the aquifer to the well, causing a lowering of the potentiometric surface.

abstracted, and is in turn drawn up by the pump. Water therefore flows from further out in the aquifer towards the borehole and so the effect continues, with the pumping causing a lowering of the potentiometric surface which spreads outwards from the borehole like a ripple from a stone dropped in a pond. We eventually reach a situation where the potentiometric surface is being steadily lowered and is sloping smoothly towards the borehole from all around in the aquifer.

When this occurs, it might seem reasonable to expect that the hydraulic gradient – the slope of the potentiometric surface – would be constant. This is not so. Actually, the shape is more like that in Figure 6.11c, with the hydraulic gradient becoming steeper as the well is approached. To see why this should be so, consider the situation shown in Figure 6.12.

In this diagram, two imaginary cylinders have been drawn, coaxial with the borehole. As water is being abstracted from the borehole at Q m^3/day, and as we have an equilibrium condition, Q m^3/day must be flowing across the surfaces of both cylinders. The outer cylinder has radius r_2, so its circumference is $2\pi r_2$ and the area of its curved surface – the area through which flow is occurring – is $2\pi r_2 b$, where b is the aquifer thickness. Similarly, the inner cylinder, radius r_1, has a curved surface of area $2\pi r_1 b$. From Darcy's law we can therefore write

$$Q = K2\pi r_1 b I_1 = K2\pi r_2 b I_2 \qquad (6.2)$$

where I_1 is the hydraulic gradient at radius r_1 and I_2 the hydraulic gradient at radius r_2. The expression $K2\pi b$ is a constant, so we can see that since r_2 is greater than r_1, I_1 must be greater than I_2 by a proportionate amount in order for the equation to balance and Q to remain unchanged. So this is the main reason why, as the water approaches the borehole, the hydraulic gradient becomes steeper, forming a characteristic lowering of the water table or potentiometric surface called the **cone of depression**.

There can be other reasons too. For the same flow to occur in the same time through a smaller cross-sectional area, not only must the hydraulic gradient become steeper but the actual speed of flow must increase. In some cases the speed may increase to the point at which the kinetic energy, and therefore the dynamic head, is significant. As we saw from Figure 6.7, this increase in kinetic energy must be compensated for by a decrease in potential energy and therefore in static head; the potentiometric surface, which indicates the static head, is therefore lowered further.

A further complication arises in unconfined aquifers. As the water table is lowered in the cone of depression, so the saturated aquifer thickness, b, is reduced. Study of equation 6.2 shows that if b is reduced,

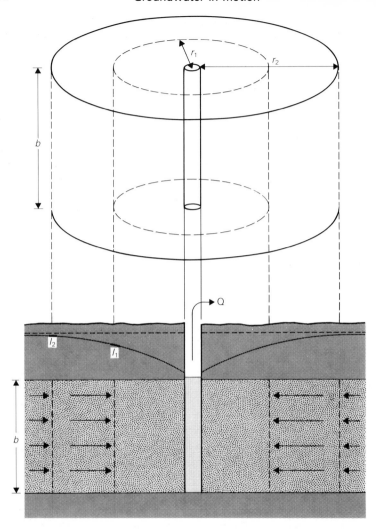

Figure 6.12 Flow to a well As water flows towards a well, water must flow through successively smaller areas at the same rate. To achieve this the hydraulic gradient must become steeper, causing the characteristic shape of the cone of depression.

I must again be increased to compensate. But if the hydraulic gradient becomes steeper, the water table in the centre of the cone of depression is lowered more, decreasing b still further so that I must be increased again. Vertical flow components and hydraulic gradients (Fig. 6.10c) also come into effect. These effects tend to reinforce each other, and this makes predicting the behaviour of wells in unconfined aquifers more difficult than the behaviour of those in confined aquifers, where the saturated thickness is constant.

Under the influence of the artificially created hydraulic gradient around the borehole, the groundwater percolates through the aquifer until it reaches the immediate vicinity of the well. Here it has one last hurdle to jump – it has to cross the well face and enter the open space of the well bore, before it can travel up or down the well to the pump. By the time it reaches the well face, the water may be travelling at such a speed that it has significant kinetic energy. On entering the well, much of this kinetic energy will be dissipated as turbulence, as the water molecules change direction in their movement towards the pump – movement which requires that some potential energy be converted into kinetic energy. Add to all this the fact that to enter the well the water may have to negotiate some form of slotted lining tube, or a portion of the aquifer whose permeability has been reduced by the act of drilling the

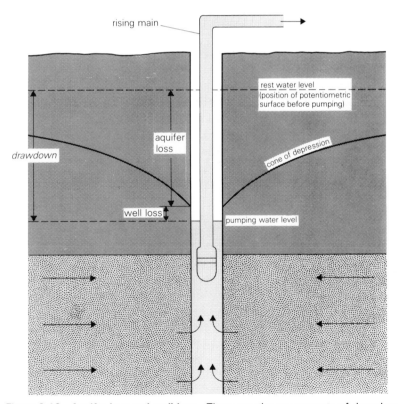

Figure 6.13 Aquifer loss and well loss These are the components of drawdown.

borehole, and the head loss involved in this last part of the journey may be a significant part of the total.

This two-part nature of the head loss causes a characteristic steepening of the cone of depression immediately around the well (Fig. 6.13). The pumping water level in the well is usually some distance below the depressed potentiometric surface just outside the well face. The total head loss (the distance between the **pumping water level** and the static or **rest water level**) is called the **drawdown**; that part of the drawdown that results from water flowing through the aquifer to the well is called the **aquifer loss**, and that part that occurs as water actually flows into the well across the well face is called the **well loss**. The ratio of aquifer loss to total drawdown is a measure of the **efficiency** of the well as an engineering structure for abstracting groundwater.

These head losses are important, because in order to lift the water to the ground surface, the pump must impart energy to the water. The greater the drawdown, the more energy is needed. This energy is not free; it has to be supplied to the pump in the form of electricity or fuel for the pump motor. To save money, therefore, it is essential that the well should be as efficient as possible − essential, in other words, that drawdown should be kept to a minimum.

A useful way of assessing the behaviour of a well is to pump the well at different rates, and measure the drawdown at each rate. The resulting pairs of figures are then plotted in the form of a yield-versus-drawdown graph, often called a **yield-depression curve**.

In a confined aquifer, provided that the pumping water level is above the top of the aquifer (so that the saturated thickness and transmissivity are constant), the yield-depression curve should in theory be a straight line, in accordance with Darcy's law that flow is proportional to hydraulic gradient. In practice, the relationship is more likely to be a curve (Fig. 6.14a).

The complication here arises from the fact that a fluid may flow in two different ways. The first way is called **laminar flow**; in this type of flow, the particles of fluid all move smoothly more or less in the same direction as the bulk of the fluid. (By fluid particles here we mean small 'boxes' of fluid, bigger than molecules but small in relation to the passageway through which the fluid is flowing.) In the other type of flow, called **turbulent flow**, the motion of the fluid particles at any point can be changing rapidly, both in speed and in direction. In laminar flow, viscous forces are dominant − in other words, the fluid is tending to resist motion of its particles relative to each other. In turbulent flow, the viscous forces are being overcome and the particles are whirling around in a much more

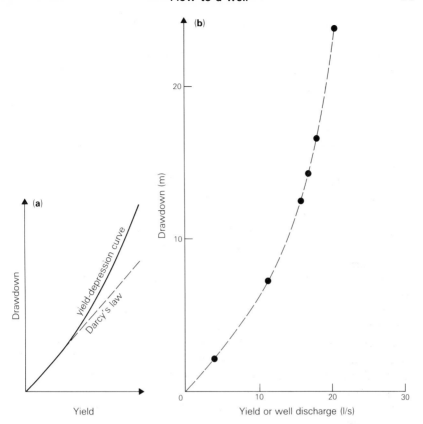

Figure 6.14 Yield–depression curves (a) Theoretical yield–depression curve for a well in a confined aquifer. If the flow through the aquifer and into the well obeyed Darcy's law, the relationship would give a straight line. But because extra energy is dissipated as the water flows into the well and up or down to the pump, the drawdown is usually greater than that predicted by Darcy's law, and the yield–drawdown relationship follows a curve. (b) Yield–depression curve for a well in the Chalk of England. The marked curvature is a result of the saturated thickness (and hence the transmissivity) of the aquifer being reduced as drawdown increases. (Based on data from the British Geological Survey.)

unruly fashion. Laminar flow typically occurs when fluid is moving very slowly through small openings (like capillary tubes) or in very thin sheets.

To express the relative importance of viscous forces in any flow condition, a number called the **Reynolds number** (symbol N_R) is used. It is calculated from the formula

$$N_R = \frac{vL}{\nu}$$

where v is the speed of flow, L is a characteristic length (usually the width of the flow passage or some obstacle in the path of the flow) and v is the kinematic viscosity of the fluid. (All of these parameters must be in consistent units, so N_R has no units — we say that it is *dimensionless*.) In the case of flow in pipes, L is the pipe diameter, and the transition from laminar flow to turbulent flow occurs when the Reynolds number is greater than about 2000.

In flow in a porous medium, an apparent velocity, Q/A, is used for v and the average diameter of the particles of rock (e.g. sand grains) is used for L. The onset of turbulence seems to occur when N_R is more than about 100, but Darcy's law appears to be valid only when N_R is less than 10. However, in most natural groundwater flow conditions, N_R is less than 1.

Darcy's law is therefore valid only for laminar flow, and may become invalid for flow speeds that are at the upper end of the laminar-flow range. At these higher speeds the flow rate may become proportional not to the head difference but to the square root of the head difference, i.e.

$$Q \propto \sqrt{h_L}.$$

Although Darcy's law is applicable to most of the flows of groundwater found in nature, the higher flow speeds that occur immediately around a well frequently result in head losses which increase in this square relationship. This means that if the well discharge Q is doubled, the head loss over this part of the flow path must be increased four times. This explains why the yield-depression curve of Figure 6.14a shows a drawdown that increases faster than the well discharge. It also explains why the designer of a well must do all he or she can to keep the speed of the water as low as possible as it enters the well. The way this is done is described in Chapter 9.

In practice it is difficult to measure direct the extra drawdown which results from well loss. What we can do, by making certain assumptions, is to calculate what the drawdown in the well would be if the only energy losses were those arising from Darcian flow in the aquifer, and compare this theoretical drawdown with that which is actually present; we attribute the difference to well losses, ignoring the fact that some non-Darcian flow may occur in the aquifer. The analysis of the two components of drawdown has resulted in many papers being published, all with the avowed (and admirable) aim of clarifying the problem, but many with the directly opposite result.

Figure 6.14b shows a real example of a yield-depression curve from a well in the English Chalk; here the increasing slope is largely a result of the transmissivity being decreased as the water table was lowered in this (unconfined) aquifer, although turbulence and vertical-flow effects are also believed to have played a part.

Summary

Groundwater moves from regions where it has high head to regions where it has lower head. **Head** is the height to which the water can raise itself above a reference level (a datum), and is measured in metres; it is a way of measuring how much energy the water possesses. Groundwater usually flows so slowly that its energy due to movement is negligible; the remaining ways in which groundwater can possess energy are by virtue of its elevation and of its pressure. When groundwater moves, some energy is dissipated and therefore a 'head loss' occurs.

When a borehole is drilled into an aquifer, the level at which the water stands in the borehole (measured with reference to a horizontal datum such as sea level) is, for most purposes, the head of water in the aquifer. Except in a few special cases the terms 'head' and 'water level in a borehole' can be regarded as meaning more or less the same thing.

Darcy's law, an empirical law discovered by Henri Darcy in a series of experiments in 1855 and 1856, describes the flow of groundwater through an aquifer. In simple terms, Darcy's law states that the flow rate Q will be directly proportional to the cross-sectional area A through which flow is occurring, and directly proportional to the **hydraulic gradient** I. The hydraulic gradient is the difference in head between two points on the flow path divided by the distance (measured along the flow direction) between them. Thus Darcy's law can be written

$$Q = KAI, \tag{6.3}$$

where K, the constant of proportionality, is called the **hydraulic conductivity**, and depends on the pore geometry of the aquifer and on the viscosity of the water.

The **transmissivity**, T, of an aquifer is the product of the hydraulic conductivity and the saturated-aquifer thickness, b.

If K is constant throughout the aquifer, the aquifer is **homogeneous**; if K is the same in all directions at any point the aquifer is **isotropic**. Darcy's law, expressed in the simple form of equation 6.3, is valid for aquifers which are homogeneous and isotropic.

When water is pumped from a well, the water level (head) in the well is lowered and a hydraulic gradient is set up towards the well from all

around in the aquifer. This causes a **cone of depression** of the potentiometric surface or water table; the reduction in head (lowering of the
water level) at the well itself is called the **drawdown**. The drawdown can
be considered to consist of two parts: an **aquifer loss**, associated with
flow in the aquifer towards the well, and a **well loss**, associated with flow
across the well face and into the well.

Selected references

Darcy, H. 1856. *Les fontaines publiques de la ville de Dijon*. Paris: Victor Dalmont. (See
    pp. 305–11.)
Fancher, G. 1956. Henry Darcy – engineer and benefactor of mankind. In *J. Petrolm
    Technol.* **8** (October), 12–14. (A brief biography of Darcy.)
Freeze, R. A. and J. A. Cherry 1979. *Groundwater*. Englewood Cliffs, NJ: Prentice-Hall.
    (See especially Ch. 2.)
Hubbert, M. K. 1940. The theory of ground-water motion. *J. Geol.* **48**, 785–944. (A classic
    and generally very readable account.)
Hubbert, M. K. 1956. Darcy's law and the field equations of the flow of underground
    fluids. *Trans. Am. Inst. Min. and Met. Engrs* **207**, 222–39.
Todd, D. K. 1980. *Groundwater hydrology*, 2nd edn. New York: Wiley. (See especially Chs
    3 & 4.)
Vallentine, H. R. 1967. *Water in the service of man*. London: Penguin. (Contains clear and
    readable explanations of water movement, viscosity, etc.)

7 More about aquifers

The term 'aquifer' was introduced in Chapter 2. A more formal definition states that an **aquifer** is a geological formation, group of formations, or part of a formation that contains sufficient saturated permeable material to yield significant quantities of water to wells and springs. This definition is by Oscar Meinzer, of the United States Geological Survey. It was apparently intended, though not explicitly stated, that the term 'aquifer' should include the *un*saturated part of the permeable material, i.e. the part above the water table, as well as the saturated part. However, we can consider that the *effective aquifer* – the part through which groundwater percolates – extends from the aquifer base to the water table.

Perched aquifers

Sometimes a layer of more or less impermeable material occurs above the water table. Infiltrating water is held up by this layer to form a saturated lens, which is usually of limited extent, above the saturated zone of the

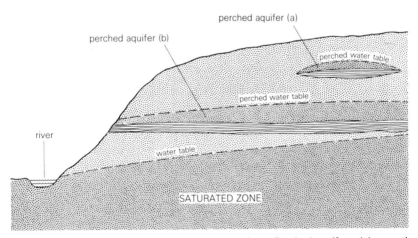

Figure 7.1 Common occurrences of perched aquifers Perched aquifers: (a) caused by impermeable material of limited extent; (b) occurring where an impermeable bed intersects a valley side some way above the river level. Situation (b) is a more common cause of perched aquifers of moderate extent than is situation (a) – in the latter, the perched aquifer will probably exist only after a period of infiltration.

aquifer proper (Fig. 7.1). An occurrence such as this is called a **perched aquifer** (and its upper limit a **perched water table**) because the ground-water in the lens is perched above the saturated zone. Perched aquifers are more common than is often supposed; although they may sometimes be only a few centimetres thick or be present only after a major infiltration event, they may in other cases be several metres thick and extend over large distances. Perched aquifers do not make large or reliable sources of supply, and it sometimes happens that the act of drilling or deepening a well penetrates the impermeable layer and allows most of the perched water to drain away.

Confined aquifers: the concept and the misconceptions

In an **unconfined** or **water-table aquifer,** the upper limit of saturation – the water table – is at atmospheric pressure. At any depth below the water table the pressure is greater than atmospheric, and at any point above the water table the pressure is less than atmospheric (Fig. 4.4).

In a **confined aquifer,** the effective aquifer thickness extends between the two impermeable layers, and at any point the water pressure is greater than atmospheric. If we drill a borehole through the confining layer, water will rise up the borehole until the column of water in the borehole is long enough to balance the pressure in the aquifer. If we imagine many boreholes drilled into the aquifer, with their water levels joined by an imaginary surface, that surface would indicate the static head in the aquifer. The term **potentiometric surface** was suggested by the US Geological Survey to replace earlier names (such as piezometric surface) for this surface; it can apply to both confined and unconfined aquifers – for example, the water table is a potentiometric surface. (Because the head may vary with depth in an aquifer – for example, as a result of recharge from above causing some vertical flow – each level in the aquifer may have its own potentiometric surface. A borehole that penetrates all or most of the aquifer, and that can receive water from all levels, measures an 'average' head and defines an 'average' potentiometric surface.)

Confined aquifers are sometimes called artesian aquifers. The term **artesian** was first applied to wells that penetrate aquifers in which the potentiometric surface is above ground level, so that on completion the well overflows or produces water without being pumped (Fig. 2.3). The term derives from the Latin name, Artesium, for the Artois region of north-west France where the phenomenon was first studied.

The classic explanation of an artesian well can be seen in Figure 7.2;

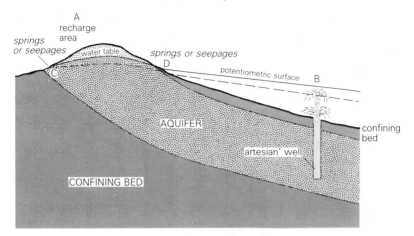

Figure 7.2   The classic explanation of an artesian well

this type of cross-section used to be common in geography textbooks. It explains the effect in simple 'water finds it own level' terms; rain falling on the aquifer outcrop at A (the **recharge area**) percolates through the aquifer to the well at B. Because of the difference in elevation between A and B, the potentiometric surface at B is above ground level. Rainfall keeps the aquifer 'topped up', the excess water being discharged by springs at C and D. If large numbers of wells are sunk at B, the discharge may exceed the replenishment. Then the potentiometric surface will be lowered (dotted line), the springs at D may cease to flow, and the wells themselves may cease to flow naturally. If the wells are then pumped, the potentiometric surface may be lowered further, below the top of the aquifer, which then ceases to be confined. This state of affairs has arisen in a number of so-called artesian basins, the London Basin – in some areas of which water in the Chalk used to be at sufficient pressure to flow to the surface, but is now as much as 90 m below ground level – being one of the best known examples.

The amount by which the potentiometric surface has been lowered below London is sometimes exaggerated, however, and a peculiar misconception has arisen about the fountains in Trafalgar Square. Three wells close to the National Gallery originally supplied water to the fountains, and to the Houses of Parliament and other government buildings. The misconception is that the fountains themselves were originally overflowing wells, with the artesian discharge making pumping unnecessary. The records show however that when the wells were completed in 1847 the water level was already 24 m below sea level. It is doubtful whether overflowing conditions ever existed in this part of London, and the misunderstanding probably arose because of a tendency in the last century to describe any well that was drilled, rather than excavated, as 'artesian'.

Confined formations which are exploited as aquifers (as opposed to deeply buried permeable layers which are too deep, or in which the water is too saline, for them to be of economic use) have a recharge area where the formation crops out at the surface. In this area the aquifer is unconfined; the same aquifer is therefore confined in one area and unconfined in another (Fig. 7.2), and it should be understood that when we speak of a 'confined aquifer' we mean an 'aquifer where confined conditions exist'; this does not mean that the formation is nowhere unconfined.

The classic explanation of confined aquifers and overflowing wells (Fig. 7.2) is reasonably accurate so far as it goes, but it is incomplete. It treats the aquifer as a simple flow conduit, conveying water from the recharge area to the wells that release the water. It ignores storage in the aquifer. It also ignores the fact that all aquifers have recharge and discharge areas which are topographically controlled (Chapter 8): flow in artesian aquifers is in many ways an extreme example of the effects found in unconfined aquifers, with the vertical hydraulic gradients increased as a result of the presence of the overlying confining bed.

It was not until the 1920s that a fuller understanding of the behaviour of confined aquifers began to be achieved. This came about because of a study of the Dakota Sandstone by Oscar Meinzer and Herbert Hard.

The Dakota Sandstone had been exploited as an artesian aquifer in North and South Dakota since 1882. There are actually two aquifers, an upper and a lower, separated by a widespread layer of lower permeability. Meinzer and Hard dealt with the upper aquifer — the one from which most of the wells drew their water — which was overlain by a sequence of shales, varying in thickness from 300 m to 500 m, acting as the confining bed.

When the first wells were drilled into this formation, some of them jetted water to heights of 30 m or more above ground. Pressures and flows were so great that water from some wells was used to drive electrical generators and other machinery. Between 1902 and 1915 there was a marked fall in pressure in the aquifer as more wells were drilled to take advantage of what many people believed to be an everlasting source of water.

By 1915, there were said to be 10 000 artesian wells in South Dakota, and in 1923 it was estimated that there were between 6000 and 8000 artesian wells in North Dakota. In one area for which there were reasonable data, the head declined by an average of nearly 4 m a year between 1902 and 1915.

Meinzer and Hard considered a strip of land, about 10 km north–south and 165 km east–west. They estimated that in the 38 years between 1886 and 1923, the wells in this strip had yielded an average combined flow of about 190 l/s or 16 000 m³/day. They also estimated the rate at which water was percolating eastward through this strip of aquifer from its recharge area in the west. They did this by making assumptions about the thickness and hydraulic conductivity of the sandstone, and from approximate knowledge of the hydraulic gradient. This calculation came out at about 25 l/s or about 2200 m³/day.

It was assumed that this east–west strip of aquifer was like those to the north and south and that all were similarly developed, so that each strip neither gained nor lost to its northerly or southerly neighbours, but was supplied with water from the recharge area to the west of it. On this basis, it was apparent that some extra source of water was contributing to the flow of these wells. The aquifer is underlain and overlain by material of low permeability, so a vertical flow of water into the aquifer was considered out of the question. In the end, Meinzer and Hard concluded that the extra water must somehow have come from storage within the aquifer itself.

Elastic storage

At first sight this may seem paradoxical. It is fairly easy to see how water is taken from storage in an *unconfined* aquifer – the water table falls and water drains from the pore space, the amount which drains being governed by the specific yield. But in the case of a confined aquifer, so long as the potentiometric surface is above the top of the aquifer, the pore space is always completely filled with water. How, then, can any water be released from storage?

For answer, think of a motor-car tyre. It is filled with air, under pressure. If you open the valve for a moment, some air comes out and the pressure drops slightly, but the tyre is *still filled* with air. Air is compressible: when we pump it into a tyre we compress it, by forcing the molecules closer together. When we open the valve, some molecules escape and the remainder move further apart. Also, the rubber tyre is elastic; it stretches and increases its volume when we pump air into it, and contracts again when we let air out.

Similar arguments apply to the aquifer. Water is compressible – not as compressible as air, but nonetheless water molecules can be squeezed a little closer together. And aquifers are elastic – the mineral grains of which they are composed can be forced apart slightly by water pressure.

Meinzer and Hard concluded that much of the water that had flowed from the wells of North Dakota had come from these sources. As water was taken from the compressible storage, the pressure in the aquifer declined, just as the pressure in the car tyre falls as air is let out.

Ironically, later work on the Dakota Sandstone has suggested that Meinzer and Hard may have been right for the wrong reason. The Dakota Sandstone used to be regarded as a classic 'artesian' aquifer, conveying water from its outcrop areas in the Black Hills and the Rocky Mountain Front Range to the wells in the Dakotas and elsewhere. Recent work suggests that the aquifer is more complex, in both its stratigraphy and its hydrogeology, than was once thought. In particular, it seems that significant quantities of water do enter the aquifer through the adjacent beds which Meinzer and Hard regarded as impermeable. Whatever the complexities of the Dakota Sandstone, however, elastic storage in confined aquifers is now an established fact.

The amounts of water that can be stored as a result of these effects are small compared with those held in the pore space, but they can nevertheless be enormous in total. We saw in Chapter 4 that if the water table in an unconfined aquifer falls by a distance z over an area A, then the volume of water that drains from the aquifer is $A \times z \times S_y$ where S_y is the specific yield. We can say that $A \times z \times S_y$ is the volume released from storage. Another way of defining specific yield would therefore be to say that it is the volume of water released from storage, in a vertical column of the aquifer with unit cross-sectional area, for each unit fall in the water table (Fig. 7.3a). Alternatively it can be considered as the volume of water taken into storage in the column with each unit rise in the water table.

In the case of a *confined* aquifer, as we have seen, the aquifer remains fully saturated (Fig. 7.3b). The weight of the overlying material is supported partly by the solid grains or framework of the aquifer, and partly by the pressure of the water in the aquifer pore space (Fig. 7.3c). When water is removed from the aquifer, the water pressure is lowered, and more weight must be taken by the aquifer framework, causing it to compress slightly. The reduction in pressure also causes a slight expansion of the water. To express the combined effect in terms of water released, we define a parameter called the **storage coefficient**, which is the volume of water released from or taken into storage per unit surface area of the aquifer for each unit change of head (Fig. 7.3b). From the wording of this definition, it can apply equally to a confined or an unconfined aquifer.

In the case of an unconfined aquifer, the volume of water that is released from or taken into storage as a result of compressibility effects

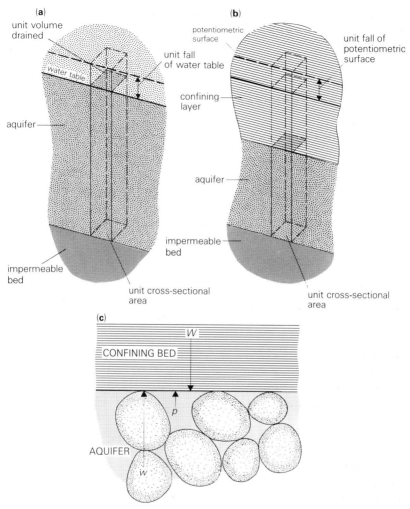

Figure 7.3 Storage concepts (a) The specific yield in an unconfined aquifer. (b) The storage coefficient in a confined aquifer. (c) The pressure (weight per unit area) W of the overlying strata is balanced partly by the pore water pressure p (measured relative to atmospheric pressure) and partly by the aquifer framework, whose *average* grain contact pressure is w, i.e. $W = w + p$. If p is reduced (e.g. by abstraction of water) then w must increase to compensate (the aquifer framework must take a greater share of the load) and the grains are pressed closer together.

(of the water and the aquifer framework) is negligible in comparison with the water involved in draining or filling of the pore space. For most purposes the storage coefficient of an unconfined aquifer can therefore be regarded as equal to the specific yield.

Fluctuations of water level

Artesian wells, spewing water to a considerable height above the ground, have an undoubted fascination. Largely because of their attraction to the authors of geography textbooks, overflowing wells are probably second only to limestone caverns as the aspect of groundwater with which most people are likely to be familiar. There are other aspects of the behaviour of confined aquifers which, though less spectacular, are equally fascinating. One of these aspects is the response of wells in confined aquifers to changes in atmospheric pressure.

Comparisons of records of water levels in such wells with records of atmospheric pressure from nearby barometers or barographs show that as pressure rises, water level falls, and vice versa. The effect is not seen in unconfined aquifers. The explanation is apparent from Figure 7.4. In the case of an unconfined aquifer (Fig. 7.4a), the water table (by definition) is at atmospheric pressure. Any change (d) in atmospheric pressure P_a can be transmitted directly to the water table both in the aquifer and in a well, and so the heads remain equal and no measurable change in water level occurs. In the case of a confined aquifer (Fig. 7.4b), the pressure W of the overlying confining bed and the pressure P_a of the atmosphere are carried partly by the aquifer framework, and partly by the water, which is of course at a pressure greater than atmospheric. Any increase in atmospheric pressure is transmitted to the top of the aquifer, and is similarly shared between the framework and the water; where a well is present, however, the pressure increase is transmitted directly to the water. The water in the well is now under more pressure (and therefore at a greater head) than the water in the aquifer; therefore some water flows from the well into the aquifer, depressing the water level in the well, until the heads of water in the well and aquifer are again equal. Such a change in water level is termed a **barometric fluctuation**. A fall in atmospheric pressure produces the opposite effect.

If the aquifer is rigid, consolidated rock, most of the increase in atmospheric pressure (transmitted as extra weight by the confining layer) will be borne by the aquifer framework, and the pressure of water in the aquifer (the pore-water pressure) will scarcely change. There will therefore be a relatively large difference between the pore-water pressure and the pressure in the well. Hence the barometric fluctuation will be much larger than in the case of an unconsolidated, weak, aquifer, where most of the increase in pressure will be transmitted to the pore water instead of being taken up by the aquifer framework, resulting in little difference in pressures between the water in the well and the pore water.

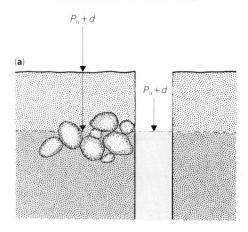

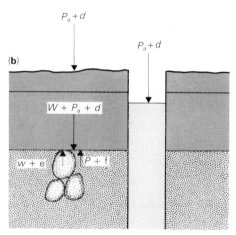

Figure 7.4 Barometric effects (a) In an unconfined aquifer any increase d in atmospheric pressure P_a is transmitted equally to water in the aquifer and to water in a well. (b) In a confined aquifer, some of the increase (e) is taken by the aquifer grains, and the remainder (f) is taken by the pore water. In the well however the full increase d is transmitted to the water surface, forcing some water from the well into the aquifer.

Atmospheric pressure changes are usually measured in millibars or in pascals, but they can be expressed in terms of head of a column of water and measured, say, in metres. Then the ratio of the change in water level in a well to the change in atmospheric pressure is called the **barometric efficiency** of the aquifer. Barometric efficiencies are often expressed as percentages. For the reasons explained above, rigid aquifers have high barometric efficiencies – perhaps as high as 70 or 80 per cent – while unconsolidated aquifers usually exhibit much lower values. Unconfined aquifers should have barometric efficiencies of zero, but there are examples of such aquifers apparently exhibiting significant barometric fluctuations. Such a fluctuation occurs when for some reason the full change in the atmospheric pressure is not transmitted directly to the pore water in the aquifer − in other words, the aquifer is not behaving as a true water-table aquifer.

Any event that causes a change in the pressure of the pore water in a confined aquifer will result in a change in water level in wells that penetrate that aquifer. This is because, at any level in the aquifer, a change in water pressure implies a change in head. This change in head, transmitted through the aquifer to the well, manifests itself as a change in water level in the well, which − as we have observed before − is essentially a manometer.

A great variety of events can and do cause pressure changes in confined aquifers. Railway trains can exert a sufficient load on an underlying confined aquifer to cause significant changes in water levels in wells. Major earthquakes can produce effects in wells several thousand kilometres away. The rise and fall of sea level during a normal tidal cycle will exert a loading and unloading effect on a confined aquifer which extends beneath the sea; the resulting changes in aquifer pore-water pressure may be observed in wells several kilometres from the shore.

The study of such water-level fluctuations can reveal important information about the aquifer. There is little doubt however that in general the most important fluctuations in groundwater level result from changes in the amount of water in storage in the aquifer. The principle can be understood by thinking of the aquifer storage in terms of a bucket. If we add water to the store, or bucket, the water level rises; if we remove water, the level falls. If we add more water than the bucket can hold, it overflows: similarly, when the water table rises above a certain level, groundwater will leave the aquifer in the form of springs or as seepage to rivers and streams (see Chapter 8).

One of the differences between the aquifer and our bucket is that the entire volume of the bucket is available to hold water, whereas most of the aquifer consists of mineral grains and water held in place by capillary forces (the specific retention); the bucket, unlike the aquifer, has a 'specific yield' of 100 per cent. Suppose the bucket to be a cylindrical and unusually large one, with a base area of 1 m^2 (Fig. 7.5a). If we pour in 100 litres of water then, since 100 litres is 0.1 m^3, the water level in the bucket will rise by 0.1 m. If we now take an identical bucket and fill it with large cobbles, arranged in such a way that they occupy 90 per cent of the volume (leaving only 10 per cent porosity), and then add 0.1 m^3

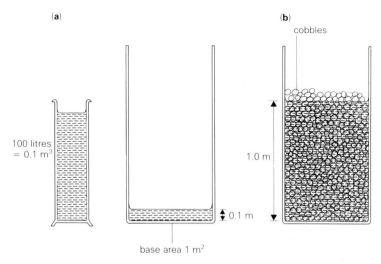

Figure 7.5 The influence of storage on water level fluctuations When 100 litres
(0.1 m^3) of water is poured into a cylindrical bucket of base area 1 m^2, the water level
rises by 0.1 m. When the same bucket is filled with cobbles to give a porosity of 10
per cent, the same volume of water causes a water-level rise of 1 m.

of water, the water level in the bucket will rise by 1 m (Fig. 7.5b). This
is because, for 0.1 m^3 of storage space to be filled, 1 m^3 of our cobble
'aquifer' has to be affected. If instead of coarse cobbles we filled the
bucket with fine sand, with a specific yield of only 1 per cent, then the
same volume of water would produce a water-level rise of 10 m. (We
should need a very tall bucket!)

If 1 mm of **effective rainfall** (that is, rainfall in excess of evapotranspiration) falls over an area, and if no infiltration or runoff occurs,
then the rainfall will result in a layer of water 1 mm deep over the ground
surface. If this layer of water then infiltrates into an aquifer with a
specific yield of 0.10 (or, in percentage terms, 10 per cent), then the water
table will rise by 10 mm; if the specific yield were 0.01, or 1 per cent, the
water table would rise by 100 mm. In general, we see that:

$$\text{rise in water table} = \frac{\text{infiltration}}{\text{specific yield (expressed as a fraction)}}.$$

Typically, in an aquifer with a low specific yield, the water table rises a
long way in response to recharge and, in the same way, falls a long way
in response to withdrawal of water from the aquifer. Aquifers with high
specific yields generally show small water-table fluctuations. Measuring
the rise in water table in response to a given amount of infiltration
therefore provides a means of estimating the specific yield.

Such estimates must be treated with caution, however. In the first place, they provide a measure of specific yield only for the zone of fluctuation; the properties of the aquifer may be different beneath this zone, so that if the water table were lowered further, perhaps by pumping, a different value of specific yield might be applicable. Second, it is rare for infiltration to occur uniformly over an area, or for the recharge to move down to arrive simultaneously at a horizontal water table. Usually, recharge will be uneven, in terms of both time and space. A large amount of rainfall may occur over one part of the aquifer during a storm; the resulting recharge will cause a temporary rise in the water table beneath that area. Immediately this rise begins, groundwater movement will take place away from that locality, so that the size of the final fluctuation in water level will depend not only on the specific yield but on the ease with which the water can be redistributed − in other words, on the transmissivity.

In the case of a confined aquifer, recharge generally occurs as a result of infiltration over an area where the aquifer is unconfined (A in Fig. 7.2). This causes a change in the position of the water table in the recharge area, and this is transmitted as a pressure change to the confined part of the aquifer, to cause a corresponding change in the position of the potentiometric surface. Withdrawal of water from the confined part of the aquifer has a direct effect. The storage coefficient of a confined aquifer is generally orders of magnitude lower than the specific yield of a water-table aquifer (a typical storage coefficient for a 30 m thick aquifer would be about 10^{-4} or 0.01 per cent), so large changes in potentiometric level can result from the withdrawal of water. Balancing this, however, is the fact that confined aquifers often extend over large areas, and the large total amount of water stored in such an aquifer may support abstraction for a long time.

Earlier in this chapter, we compared a confined aquifer to a motor-car tyre. A car tyre stretches and expands slightly when inflated, and contracts correspondingly when deflated, and it was implied that a confined aquifer does the same. If this is so then the withdrawal of large volumes of water from confined aquifers at rates much greater than natural replenishment should lead to a contraction of the aquifers and a corresponding subsidence of the land surface. The occurrence of such subsidence is powerful evidence for the elastic behaviour of confined aquifers. Notable examples of such subsidence have occurred in the San Joaquin Valley of California, and around Mexico City; in both cases, the ground surface has subsided by several metres. In Mexico City, in particular, considerable damage has occurred to buildings. Pumping from aquifers beneath Venice produced subsidence of about 0.2 m which, given the location of the city and other factors, has proved to be very serious.

Such spectacular effects, although serious, are comparatively rare; they tend to occur in thick deposits containing fine sands, silts and clays, and in many cases are the result of excessive pumping from aquifers. Unfortunately, it is often assumed that subsidence is an inevitable consequence of groundwater abstraction, and local people sometimes oppose the development of groundwater-abstraction schemes under the impression that these will lead to subsidence on the same scale as that associated with coal mining. Such fears often stem from false concepts of groundwater occurrence, such as the ideas of underground rivers and lakes mentioned in Chapter 2.

Rock types as aquifers

Several factors can provide a basis for classifying rocks – for example chemical composition, mineral composition, age and texture – but, as a starting point, it is usual to group all rocks into three main types, depending on their origin. These are *sedimentary, igneous* and *metamorphic,* and most of the world's major aquifers are of sedimentary origin. Igneous and metamorphic rocks, in general, are far less important as sources of groundwater. To see why, it is only necessary to consider how the three rock types form.

Igneous rocks
Igneous rocks form by the cooling and solidification of molten rock or **magma,** which may be forced into other rocks, forming an **igneous intrusion,** or which may be **extruded** at the surface (as for example in a volcanic eruption). This method of formation means that there are usually few voids in the rock at the time of its formation – perhaps just a few, small, unconnected cavities or **vesicles** caused by the presence of bubbles of gas. The rock therefore has very little initial porosity or permeability; only in the case of extrusive igneous rocks (lava) are there likely to be large or interconnected openings (p. 86). Basalt lavas, in particular, are well known for their tendency to form columns separated by sizeable cracks or joints as the lava cools and contracts.

Later in the life of a rock, porosity and permeability may increase. Weathering may weaken and remove some minerals to create voids or to open up joints; tension resulting from movements in the Earth's crust, or stress release as the weight of overlying rock is removed by erosion, may cause fractures to develop and open. To distinguish between voids present when the rock was formed and those which develop as a result of later processes, some workers use the terms **primary porosity** and **secondary porosity**.

Metamorphic rocks

Metamorphic rocks are formed by the alteration of other rocks under the action of heat or pressure. Small occurrences may be due to the baking of other rocks by hot magma, in which case existing porosity may be preserved, but large occurrences of metamorphic rocks are the result of processes deep within the Earth's crust. The temperatures and pressures involved mean that these rocks are altered and compressed to such an extent that voids are destroyed. Only when these rocks are brought back to the surface, as overlying rocks are removed by erosion, is there a chance of secondary porosity developing in the same way as for igneous rocks.

Sedimentary rocks

Sedimentary rocks form as a result of deposition of particles, which are often derived from the weathering and erosion of other rocks. This deposition usually takes place under water, frequently on the sea bed, but it may occur in river beds or lakes, or even on dry land. The nature of the process means that particles will be deposited with spaces between them; the size of these voids will depend on the sizes of the particles, and on how well sorted they are. Clearly, small grains will have small pores between them; large grains, all of one size, will have large pores. If the sediment contains large and small grains – i.e. if it is 'poorly sorted' – the small grains will tend to occupy the voids between the larger ones, leading to a lower porosity than in the case of well-sorted sediments.

Fine sediments, such as clays and silts or fine sandstones, may have high porosities, but the pores are so small that surface tension or molecular forces prevent water movement, so that in these materials permeability is low. Coarse sands and gravels, especially if they are well sorted, are very permeable.

After deposition, water percolating through the sediment may deposit material brought in solution from elsewhere. This deposition on and around the mineral grains is termed 'cement', and it binds the sediment together. The process may be assisted by compaction resulting from burial beneath other layers of sediment. In this way a **non-indurated sediment** or **unconsolidated sediment** becomes an **indurated sediment** or **consolidated sediment**. At the same time, porosity is reduced; in extreme cases the cement may fill up almost the whole of the primary porosity.

Once a sediment has become consolidated it can be subject to fracturing and the development of secondary porosity in the same way as igneous or metamorphic rocks.

UK aquifers

The United Kingdom has a varied and interesting geology, a fact that explains the diversity of scenery within a small area. In Britain, in very general terms, older rocks occur in the north and west and younger rocks in the south and east (Fig. 7.6). The older rocks which form most of Scotland, Wales and Northern Ireland are mainly metamorphic and igneous rocks or well-indurated sedimentary rocks which have low permeabilities, so these areas have little in the way of useful aquifers. In Scotland and Northern Ireland, groundwater is used to provide about 5 per cent of public water supplies, while in England and Wales, taken together, about one-third of the water used for public supply is provided by groundwater.

In other countries things may be different. Sweden, for example, is composed almost entirely of igneous and metamorphic rocks of low permeability, and obtains only a few per cent of its water supplies from underground. In Denmark, which is composed entirely of sediments, the situation is almost exactly reversed – 98 per cent of water is drawn from aquifers, and only about two per cent from surface sources.

Total groundwater abstraction in England and Wales is about 2400 million cubic metres (or 2.4×10^{12} litres) per year. That is enough water in a year to cover the whole of England and Wales to a depth of about 15 mm, or to supply every man, woman and child in England and Wales with 49 000 litres of water each year (or 135 litres/day).

The main aquifers in England and Wales consist of partially indurated sedimentary rocks. In terms of the quantity of water abstracted, the Chalk is the most important, followed by sandstones of the Permian and Triassic Systems. In 1977 abstraction from the Chalk accounted for more than 50 per cent, and from the Permo-Triassic sandstones about 25 per cent, of total groundwater abstraction in England and Wales. These two aquifer groups therefore deserve consideration in some detail.

Chalk
Chalk is a soft white limestone. Limestones are composed of calcium carbonate, and they usually consist of the skeletal remains of aquatic organisms – shell fragments are a common source. The Chalk of Britain

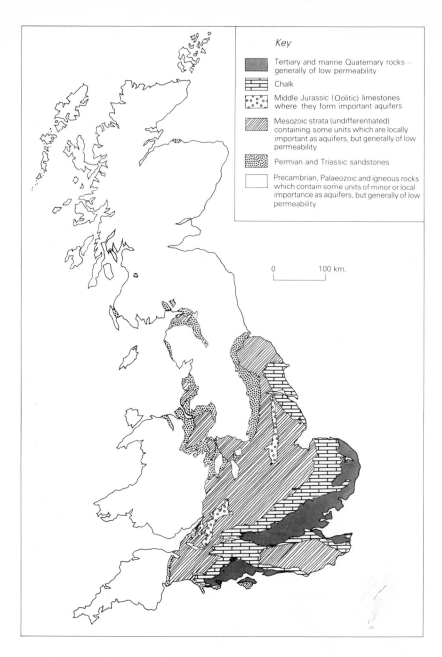

Figure 7.6 Simplified geological map of Great Britain, showing the most important aquifer groups (Minor outcrops omitted.)

was formed in the Upper Cretaceous period of geological time, and consists of shell fragments and foraminifera with sizes typically between 10^{-2} and 10^{-1} mm (10 to 100 μm), set in a finer matrix. The matrix is so fine that it was at one time believed to be of inorganic origin, perhaps a chemical precipitate. It was not until the availability of the electron microscope in the 1950s that this finer material was also shown conclusively to be of organic origin, consisting of whole or broken minute calcareous shells (coccoliths) of plankton (Fig. 2.1b).

At the time of deposition, the Chalk probably had a porosity of at least 50 per cent, the high value being partly due to the hollow shells. In southern England there has been limited cementation, and porosities of 40 per cent are still common throughout much of the aquifer; in Yorkshire and Lincolnshire, cementation has been more extensive, reducing porosities to 10 to 20 per cent. In Northern Ireland, the process has gone further; in the Chalk here, known as the 'White Limestone', porosities of less than 5 per cent are common.

Most of the Chalk is a pure white limestone containing lumps or layers of flint, but in the lower part of the formation clay minerals and marl bands are common, imparting a greyish colour. Flints are absent from this lower part.

The fine-grained nature of the Chalk means that the pores and pore necks are correspondingly small, the latter typically less than 1 μm. Most pore water is therefore held virtually immobile by capillary forces: in spite of the high porosity, specific yield is therefore low – generally about 1 per cent. Permeability is also low if measurements are made on small samples, which take into account only the intergranular permeability. (The example in Figure 2.1b, from a borehole on the Berkshire/Hampshire border, has a measured porosity of 42 per cent and a hydraulic conductivity of 0.003 m/day.) Typical hydraulic conductivity values measured in the laboratory on Chalk samples are between 10^{-3} m/day and 10^{-2} m/day; given representative thicknesses for the Chalk of 200 m to 500 m, this would imply transmissivities of less than 5 m^2/day. In practice, measurements of transmissivity from wells in the Chalk give values that frequently exceed 1000 m^2/day. The difference is caused by the presence of cracks or fissures; these are usually in three directions, more or less mutually perpendicular, one set being approximately parallel to the bedding. These have quite limited openings (usually less than 1 mm) except where they have been enlarged by the dissolution of calcium carbonate, when they may have openings of several millimetres (Fig. 7.7).

The Chalk is thus an unusual aquifer. Fissures may contribute about 1 per cent porosity, but almost the whole of the specific yield and more

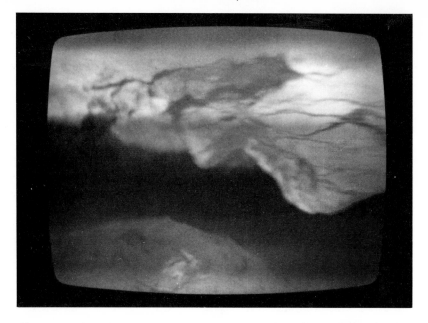

Figure 7.7 Fissure permeability in the Chalk Photograph taken using a closed-circuit television camera of the wall of a Chalk borehole, showing a fissure that has been enlarged by solution. The view is of an area about 80 mm × 60 mm. (Photograph published by permission of the Director, British Geological Survey.)

than 99 per cent of the transmissivity. Intergranular pore space contributes a porosity of 20 to 50 per cent of bulk volume, but very little to specific yield or transmissivity. At great depths, fissures tend to be closed by pressure. Further, at significant depths below sea level there is little or no circulation of fresh water to enlarge fissures by dissolution. In most areas, therefore, it is the upper part of the Chalk that is the effective aquifer. Under these circumstances and with perhaps one or two fissures contributing 80 per cent of the transmissivity, formulae like

$$T = Kb$$

(p. 54) cease to be of practical significance.

Because of the high transmissivity of the Chalk, hydraulic gradients are usually low and the water table is correspondingly flat. Thus in areas with marked relief the water table is often many metres below ground level, giving rise to a thick unsaturated zone. The small pore sizes and corresponding high specific retention mean that this unsaturated zone is almost fully saturated – only fissures and a few large pores drain under gravity. However, the pore sizes are such that a great deal of the water can be removed by plants, and there seems little doubt that in dry weather water can move up in limited quantities from depths of a few metres in response to evapotranspiration losses.

Permo-Triassic sandstones

In contrast to the Chalk, which is a marine deposit, the sandstones of the Permian and Triassic Systems in Britain were laid down at a time when most of the country was desert or semi-desert. Some of the sandstones appear to have originated as sand dunes, which have subsequently become partially cemented; the majority were deposited by apparently ephemeral rivers and lakes in a semi-arid environment. Deserts are not renowned for an abundance of vegetation or wildlife, and hence organic remains – the fossils geologists use as a major means of comparing the ages of rocks – are rare in these sandstones. It is therefore difficult to decide whether some of the deposits are Permian or Triassic, and they are often grouped together as Permo-Triassic.

In addition to sandstones, the Permo-Triassic system includes conglomerates, siltstones and mudstones. The nature of their deposition means that different materials were being deposited in different places at the same time, so that there is a great deal of lateral variation in the rocks.

In general, the Permo-Triassic sandstones form a much more 'normal' aquifer than the Chalk. Porosity depends on the degree of sorting and rounding, the packing and cementation and so is highly variable, but values of 20 to 35 per cent are usual; specific yield, which is controlled by grain size as well as the other factors, is typically 15 to 25 per cent, although in practice variations in pore size throughout the aquifer may inhibit drainage of the pore space. Hydraulic conductivity values are typically between 1 m/day and 10 m/day in the coarser sandstones, and between 10^{-1} m/day and 1 m/day in the finer deposits. Desert sandstones, deposited as sand dunes, generally contain rounded grains (Fig. 2.1a) which have been rolled by wind action; because finer material is blown away, they are also well sorted. These round grains cannot pack tightly, so such sandstones (e.g. the Penrith Sandstone of the Eden Valley) form some of the most permeable of British aquifers, with hydraulic conductivity values as high as 20 m/day. (The example in Figure 2.1a, from the Permian sandstone of the Eden Valley, Cumbria, has a porosity of 31 per cent and a hydraulic conductivity of 7.7 m/day.)

There is evidence that fissures are important locally in these sandstones and may play a major part in allowing water to flow easily into wells, but they may not carry water on a regional scale in the same way as in the Chalk. Because the intergranular permeability is relatively high, it is easier for water to move from intergranular pore space into fissures, or vice versa, than is the case in the Chalk. The relatively high specific yield means that water-table fluctuations are usually much smaller than in the Chalk.

Other aquifers – unconsolidated sediments

In the United Kingdom, unconsolidated sediments are of minor importance as aquifers, but in other parts of the world they may be the main or only source of groundwater and they provide the bulk of the world's developed aquifers. They are generally of relatively recent origin, and lack of compaction and cementation means that they include some of the most permeable natural materials. Of particular importance are sands and gravels in river valleys, which are usually very permeable but of limited thickness and extent unless the river is a major one; the small size of UK rivers means that alluvial aquifers here are of only local importance.

Where the valley is controlled by geological faulting, so that the deposits are filling a trough of structural as well as erosional origin, the deposits may reach thicknesses of hundreds or even thousands of metres; the central valley of California is an example. In other cases the deposition of sediment takes place over a larger area but with smaller thicknesses. The groundwater of these broad valleys is generally a more important resource than that of the deeper fault-controlled valleys; not only do the broader valleys contain a greater total volume of sediment, but the water is frequently of better quality, as the very deep deposits of the rift-type of valley may contain saline water. The groundwater of the Ganges delta, for example, is vital to the peoples of India and Bangladesh.

Unconsolidated sediments were probably the first aquifers to be developed. River alluvium, in particular, was probably an obvious choice for early wells, offering ease of excavation, a shallow water table and a demonstrable connection with surface water. Similar considerations mean that these deposits are still heavily utilised today in the industrialised as well as in the developing world. Historically, the kanats (long underground galleries, collecting groundwater from detrital deposits along the foothills of the mountains and conveying it to the cities of the plains) of Persia and neighbouring countries are among the oldest known public waterworks (Chapter 9).

Non-aquifers

Not all the rocks in the Earth's crust are aquifers. At great depths, conditions are such that open voids cannot exist, and rocks therefore have zero porosity. Before these depths are reached, conditions generally exist under which pores and fractures are so reduced in size that, although the rocks may have measurable porosity, for practical purposes they are impermeable. These rocks thus serve as the lower boundaries of the deepest aquifers.

At lesser depths, there may be rock formations that have aquifers above or beneath them but which themselves have permeabilities too low for them to be called aquifers; a confining bed is an example of such a formation. These formations frequently contain water – that is to say they are porous – but do not allow water to move through them under typical hydraulic gradients; such formations are sometimes called **aquicludes.** Other formations permit water to move through them, but at much lower rates than through the adjacent aquifers; in particular they may permit the vertical flow of water between underlying and overlying aquifers. There was a suggestion that this type of formation be called an **aquitard.** In 1972, however, the United States Geological Survey published recommendations which sought to end some of the confusion that was arising over the use (or misuse) of hydrogeological terms – an act for which the USGS deserves the gratitude of all English-speaking hydrogeologists. One of the recommendations was that terms like 'aquiclude' and 'aquitard' should be discontinued, and reference be made simply to confining beds with some description of their permeability relative to the adjacent aquifer. Whether or not this recommendation eventually wins support, terms like 'aquiclude' and 'aquitard' are common in existing literature, so it is as well to know what they mean.

These formations of low permeability are important. True, they will not yield water to wells, but they play an important part in controlling the movement of water in adjacent permeable formations. Furthermore, as we shall see in Chapter 12, there are times when the hydrogeologist or engineer deliberately seeks impermeable rocks – as suitable material in which to make a deep excavation, for example, or as a suitable location for a dam.

Igneous and metamorphic rocks

The principal difficulty facing anyone dealing with the hydrogeology of igneous or metamorphic rocks is their extreme variability. This is largely because these rocks possess little primary porosity: in general the porosities of unweathered pieces of metamorphic or intrusive igneous rocks are less than 1 per cent, and hydraulic conductivities of such pieces are unlikely to exceed 10^{-5} m/day. The water-bearing capacity of the rocks is therefore related almost entirely to the secondary porosity, which develops as a result of fracturing or weathering. The effects of weathering are usually limited to depths of less than 100 m, and below this fractures tend to be closed by the weight of the overlying rock. The permeability of these rocks therefore decreases with depth. There are always exceptions; mines in the Canadian Shield, for example, have recorded significant inflows more than 1000 m below the surface.

Because the extent of weathering and fracturing varies from one rock type to another, and varies with geological history and climate within the same rock type, it is almost impossible to generalise on the hydro-geological properties of these rocks. Large variations within a single rock type are possible over small areas, so that detailed investigations are necessary before any predictions can be made as to well yields or the amount of water likely to enter an excavation. Unfortunately, the cost of detailed investigations can only rarely be justified – more usually in assessing the difficulties of disposing of water from underground works than in predicting well yields for water supply.

Extrusive rocks tend to be even more variable than the intrusives. The dense varieties have primary porosities of less than 1 per cent, while pumice may have porosities as high as 85 per cent. Pumice is essentially solidified foam, consisting of lava containing numerous vesicles. Because the vesicles are rarely interconnected, the permeability of pumice is usually low.

Many hollow, generally tubular structures can form in lavas, but where high permeability exists in lavas it usually results from one of two causes: the spaces that form as a result of shrinkage into columnar struc-tures, and the voids that occur between successive lava flows as one is deposited upon the weathered surface of its predecessor.

The weathering of lava flows leads to the formation of secondary minerals which can fill the primary pores. In many cases, therefore, the permeability of these rocks decreases with time. Some lavas (such as the basalts of the Snake River and Columbia areas of the United States) form important aquifers, while others, such as the Deccan basalts of India, have low permeabilities.

Selected references

Barrow, G. and L. J. Wills 1913. Records of London wells. *Memoirs of the Geological Survey of England and Wales.* London: HMSO.

Davis, S. N. and R. J. M. De Wiest 1966. *Hydrogeology.* New York: Wiley. (Chs 9, 10 and 11 provide a useful and straightforward account of various rock types as aquifers.)

Helgesen, J. O., D. G. Jorgensen, D. B. Leonard and D. C. Signor 1982. Regional study of the Dakota aquifer (Darton's Dakota revisited). *Ground Water* **20** (4), 410–14.

Lohman, S. W. *et al.* 1972. *Definitions of selected ground-water terms – revisions and conceptual refinements.* US Geological Survey Water-Supply Paper 1988. Washington, DC: US Govt Printing Office.

Meinzer, O. E. and H. A. Hard 1925. *The artesian water supply of the Dakota Sandstone in N. Dakota, with special reference to the Edgeley Quadrangle.* US Geological Survey Water-Supply Paper 520-E. Washington, DC: US Govt Printing Office.

Rodda, J. C., R. A. Downing and F. M. Law 1976. *Systematic hydrology.* London: Newnes-Butterworths. (Ch. 5 contains an excellent and concise account of the major British aquifers.)

Todd, D. K. 1980. *Groundwater hydrology,* 2nd edn. New York: Wiley. (See especially Ch. 6.)

8 Springs and rivers, deserts and droughts

Discharge from aquifers

Earlier chapters have shown something of how water enters an aquifer, how it is stored there and the factors that govern its movement through the aquifer. Now it is time to consider how water leaves the aquifer. One way it can leave is by abstraction from a well, and we shall see more of this in Chapter 9. But water was entering and leaving aquifers long before there were people on Earth to sink wells, so let us first look at how nature accomplishes this part of the water cycle.

Water that leaves aquifers naturally usually finds its way into river systems (although water from coastal aquifers may discharge direct to the sea). The area of land that drains to a river is called the river's **catchment** or 'catchment area'; the higher land that separates two catchments is called, in Britain a **watershed** or a **divide**. (In the United States, where the term 'watershed' is often used for a catchment area, a true watershed is always called a 'divide'). In impermeable terrain, the catchment contributes overland flow and interflow; where permeable rocks crop out within the catchment the river will usually derive some of its flow from groundwater. In the latter case, the river can be considered to have a **surface-water catchment** and a **groundwater catchment**; if the water-table relief exactly follows the ground relief, the two divides will coincide, as in Figure 8.1a. This will occur if the rocks are homogeneous and isotropic, and may occur even if they are not. The groundwater flow paths for a homogeneous, isotropic catchment take the form shown in Figure 8.1b. Flow occurs from the **recharge areas**, which are on high ground, to the **discharge areas**, which are low-lying. Note that flow occurs throughout the aquifer; there is no stagnant zone at depth.

Figure 8.1 Recharge and discharge areas   (a) Surface-water divides (S) and groundwater divides (G) between valleys in a region composed of permeable, homogeneous and isotropic rock. (b) Flow pattern of groundwater in the catchments shown in (a). (c) Vertical components of groundwater flow imply vertical hydraulic gradients which can be measured using piezometers. Note that head decreases with depth below recharge areas (R) and increases with depth below discharge areas (D). These effects are shown by the water levels in the piezometers at R and D. In the deep piezometer at D, the head is above ground level.

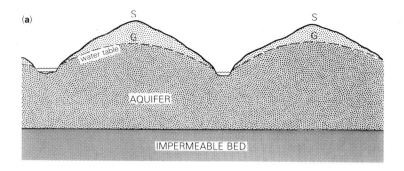

(a)

S

G

water table

AQUIFER

IMPERMEABLE BED

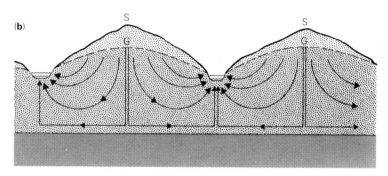

(b)

S

G

S

G

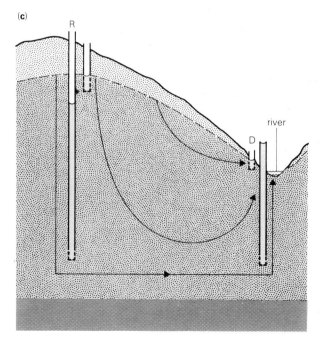

(c)

R

river

D

The vertical-flow components mean that there must be vertical components of hydraulic gradient. The resulting head differences between points at different levels in the aquifer can be measured using piezometers (p. 55). There is a downward component of flow (and therefore of hydraulic gradient) under recharge areas and an upward component under discharge areas (Fig. 8.1c). These head differences occur even if the aquifer is perfectly homogeneous and isotropic; they may be increased by the variations of permeability with depth which are common in many aquifers. The classic 'artesian' situation (Figs 7.2 and 8.5) is an extreme case of the condition in the deep piezometer at D (Fig. 8.1c).

The groundwater and surface-water divides rarely coincide as precisely as shown in Figure 8.1a. Sometimes major departures occur, as shown in the theoretical case of Figure 8.2, where the permeable bed B is effectively transferring water from the surface-water catchment area of river C to that of river D; as a result, the groundwater divide G is closer to C than is the surface-water divide S. The groundwater catchment of the River Itchen, in the Chalk of Hampshire, for example, is estimated by Ineson and Downing to be some 20 per cent larger than the surface catchment. The most extreme examples of non-coincidence of groundwater and surface-water divides occur in cavernous limestone areas.

Water will flow from an unconfined aquifer wherever the water table intersects the ground surface (Fig. 8.3). Where the flow from an aquifer is diffuse it is termed a **seepage**; where it is localised, as for example along a fault or fissure (Fig. 8.3a) it is called a **spring**. It is common to find lines

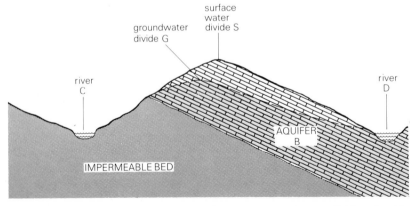

Figure 8.2 Non-coincidence of surface-water and groundwater divides In this theoretical example, transfer of groundwater, through bed B from the catchment of River C to the catchment of River D leads to the non-coincidence of the surface and groundwater divides.

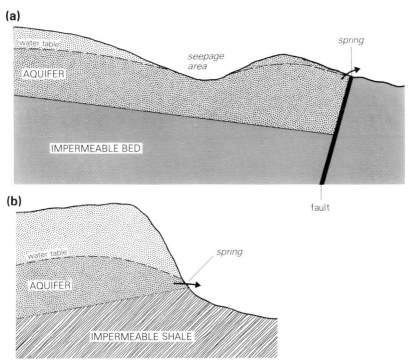

Figure 8.3   Common occurrences of springs and seepage areas

of springs or seepages where permeable sandstones or limestones form
high ground and rest on less permeable rocks such as shales or clays (Fig.
8.3b); these spring lines are often used by field geologists as a guide when
mapping the boundary between two such formations.

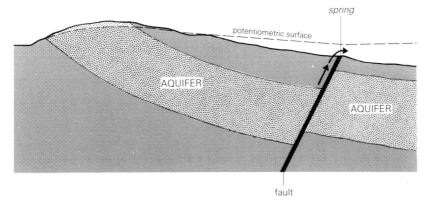

Figure 8.4   Discharge from a confined aquifer   A fault plane can provide a permeable
path along which groundwater can discharge from a confined aquifer.

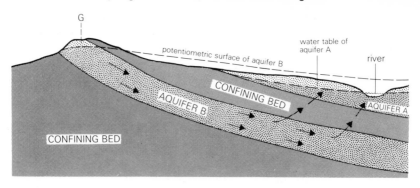

Figure 8.5 Discharge from a confined aquifer   Natural discharge from a confined aquifer (B) across a confining bed. If there were no discharge from B, its potentiometric surface would be horizontal from G eastwards.

The largest known spring in the world issues from a limestone at Ras-el-Ain in northern Syria. Its flow, at a rate of about 40 m³/s, helps to sustain the flow of the Euphrates via its tributary, the Khabour.

Water will flow from a confined aquifer where the potentiometric surface is above ground level *and* where there is locally some form of permeable path through the overlying confining bed (Fig. 8.4). However, it is much more common for groundwater to leave a confined aquifer by percolating slowly through a confining bed into permeable material (Fig. 8.5). This is one of the mechanisms ignored by the classical explanation of artesian aquifers (pp. 67–8).

Why rivers keep flowing

As Figure 8.1b shows, discharge of groundwater occurs where the bottom of a river valley lies below the water table. The discharge may take place through the bed or bank of the stream or river and so may not be visible, but such discharges account for the greatest proportion of flow from aquifers. A river that receives water from an aquifer, like that in Figure 8.1b, is termed a **gaining stream**. For a river to flow throughout the year, even during long periods without rainfall, it must have a source of water other than surface runoff or interflow. This water, which sustains the river throughout dry weather, is present, though less apparent, at other times; it is termed **baseflow**. Baseflow can be provided by groundwater discharge from an aquifer, from surface-water storage (as in the case of a river that flows from or through a lake) or from the melting of glacier ice or of snow which is present throughout most of the year. The first of these sources is the most common, and many writers use the terms 'baseflow' and 'groundwater discharge' as though they were synonymous.

It may be that a stream flows across permeable material but that the bed of the stream is higher than the water table. In such a case, unless the stream bed is itself impermeable (perhaps floored with clay, for example) water will flow from the stream to the aquifer; in this case, the stream is called a **losing stream**. It is possible for a river to be a gaining stream over one part of its length and a losing stream over another part; or for the same stretch to be a gaining stream at some times and a losing stream at others, as the water table rises and falls.

A river that flows throughout the year, every year, is called a **perennial stream**, and a river that flows only occasionally, perhaps for hours or days in several years, is called an **ephemeral stream**. Ephemeral streams usually occur in semi-desert regions, where rainfall, though unpredictable and unusual, may be heavy and localised in the form of storms. A sudden downpour produces sufficient surface flow or interflow to sustain the stream for a short time, before its flow is lost by evaporation or infiltration. Ephemeral streams rarely have well-defined channels and are never gaining streams.

Finally there are **intermittent streams**. These flow for part of each year, usually during or after the season of most rainfall, or as a result of snowmelt. One occurrence is where the water table, after infiltration in the winter, rises above the bed of the upper reach of a river (Fig. 8.6); as the height of the water table declines, the source of the river travels down from A to B. Downstream of B, which is termed the **perennial head**, the stream is perennial. The situation shown in Figure 8.6 is common in the Chalk downlands of southern England where, because of the low specific yield of the Chalk, large fluctuations occur between the spring and autumn positions of the water table. These intermittent

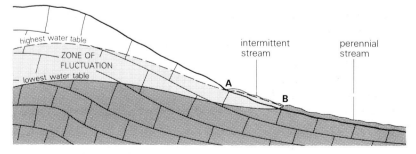

Figure 8.6   An intermittent stream   An intermittent stream may occur in response to a changing water table.

portions of Chalk streams are called **bournes**, which explains the common occurrence of 'bourne' or 'winterbourne' in the place names of the
area.

A common statement is that groundwater supplies about one-third of
all the water used for public supplies in England and Wales; by 'groundwater' the authors mean water pumped from wells or collected from
springs. The statement is perfectly true – it is made in this book – yet
it underestimates the importance of groundwater in supplying the needs
of England and Wales, because it ignores the contribution that groundwater makes to the flows of our major rivers, many of which are used
as sources of supply. As we have just seen, rivers can only flow
throughout the year, even in a temperate climate like that of Britain, if
they have a source of baseflow. In Britain, the baseflow component of
all our major rivers is derived from groundwater. It therefore follows
that, during dry periods, the water abstracted for public supply from
rivers is indirectly derived from aquifers. To estimate the importance and
quantity of this baseflow it is necessary to know something of how river
flows are measured and analysed.

Measuring river flows

The **discharge** of a river or stream is the volume of water flowing past
a given point in a unit of time; it is therefore the cross-sectional area of
the flow section multiplied by the speed with which the water is flowing.
If the shape of the river channel is known, then the cross-sectional area

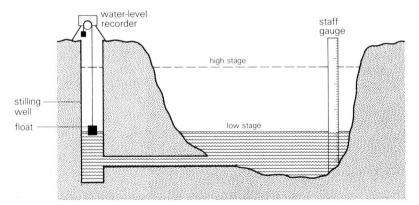

Figure 8.7 Cross-section of a river channel at a gauging station The stilling well
eliminates ripples and wind effects, to provide a smooth surface for water-level
measurement.

can be determined at any time provided that the depth of water in the river is known at that time (Fig. 8.7). The depth of water is usually measured in terms of the height of the water surface above a reference or datum level; this height is called the river **stage**, and a graph of river stage against time is called a **stage hydrograph**. Stage is most simply measured using a vertical graduated post (a **staff gauge**) set in the river bed or against the bank; these staff gauges can be seen at intervals along most British rivers. More usefully, recorders can be installed which measure water level continuously and which produce a hydrograph automatically (Fig. 8.7). The datum from which the stage is measured is not necessarily the bed of the river at that point; often stage measurements at several points along a river are made relative to a common datum.

Stage measurements are of importance to engineers concerned with river management, for purposes such as navigation and for flood prediction and control. Those – including hydrogeologists – whose primary concern is water resources are more interested in the river discharge, for which it is necessary to know the speed of flow as well as the stage.

Flow speed can be measured in a variety of ways but it is of little use to know the speed and the stage – and therefore the discharge – at only one time. Much more valuable is the record of flow plotted against time which is depicted in a **discharge hydrograph**. As we have seen, it is relatively easy to measure the river stage and to produce a stage hydrograph. If the hydrologist or engineer can produce a graph or formula that relates discharge to stage for the entire range of flows which may occur at that point on the river, then subsequently he can derive the discharge value whenever he wants simply by measuring river stage. The point on the river where this is done is called a **gauging station**, and Figure 8.8 shows the form of a **stage–discharge** relationship or **rating curve** for such a station.

There are two principal ways of establishing the stage–discharge relationship. The first is by choosing a suitable stream section, and then measuring the cross-sectional area and speed at various stream stages and so constructing a rating curve like that in Figure 8.8. A gauging station that uses this method is a **velocity–area** station. A simple (though not very accurate) way of measuring flow speed is to drop floats into the water and time them over a known distance. A more accurate and common method is to use a **current meter**. This is an instrument with a rotating propeller or anemometer-like vane, which is mounted on a rod or on a weighted cable so that it can be lowered into the river flow to any desired depth. The rod or cable can simultaneously be used to measure the depth of the stream; by making measurements of speed and depth

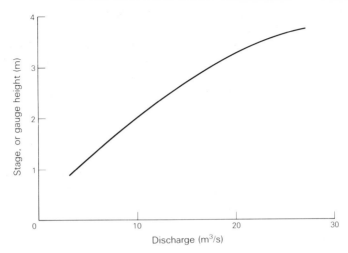

Figure 8.8   A stage-discharge graph   The form of a stage-discharge graph or rating curve for a river gauging station.

across the river section, the rate of flow can be calculated for that water level or stage. In practice, measurements may have to be made at several depths at each point as the speed usually varies with depth. On very large rivers measurements may be made from boats or by lowering the meter from a specially constructed cableway. And all the measurements must be repeated, a great many times, so that the full rating curve can be produced.

At some modern gauging stations, current meters are not used. Instead ultrasonic signals are sent diagonally across the river, underwater, between special transmitters and receivers. Going in one direction the signals are speeded up by the movement of the water, and in the other direction they are slowed; the difference between the two sets of travel times enables the flow speed to be calculated.

The problem with all of these velocity–area stations is that the relationship between stage and discharge changes with time, as a result of river bed or bank erosion, growth of aquatic plants and other factors. The second way of establishing the stage–discharge relationship avoids many of these problems by building an artificial structure – a weir or a flume – across the river. A **weir** is essentially a wall over which the water flows, while a **flume** is essentially a throat – a reduction in width or depth, or both – through which the water flows with increased speed. For both weirs and flumes, provided that the structure is built in a standard way and that its dimensions are known, the flow through or over it can be calculated from formulae if the height of water is known. As the height is known from the river stage measurement, the latter can be converted directly to a discharge value.

For reasons of cost these structures cannot be built on large rivers, for which velocity–area stations remain the only practical means of measuring discharge. Flumes tend to be more expensive than weirs; they are favoured on small upland streams, particularly those that carry a lot of silt or debris. This material can build up behind a weir and change its characteristics, whereas the increased speed of the water as it passes through the reduced cross-section of the flume tends to keep the structure clear.

It should be noted that many weirs are built primarily for purposes other than flow measurement. On larger rivers, these purposes include prevention of flooding by controlling the rate at which water is allowed to flow from one stretch of the river to the stretch downstream, and the maintenance of sufficient depth of water for navigation.

Hydrograph analysis

Having gone to such lengths to collect this streamflow information, how can we use it? Figure 8.9 shows a theoretical stream-discharge hydrograph for a river draining a catchment that is underlain by some permeable and some relatively impermeable material. A few years ago, a classic exercise for hydrologists or hydrogeologists would have been to

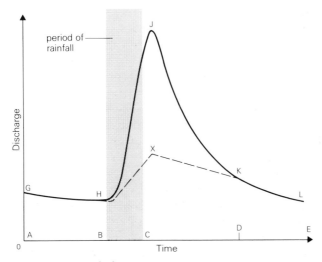

Figure 8.9 Hydrograph analysis

analyse such a hydrograph, a process which involved deciding how much of the flow at various times was derived from groundwater flow, how much from surface runoff and how much from interflow. The theories behind this analysis were based largely on the work of R. E. Horton in the USA, who, in the 1930s, put forward the idea of infiltration capacity being a major control on the way rainfall is disposed of over a river catchment. With the realisation that overland flow and surface runoff were rare events in many catchments, objections to Horton's description of streamflow began to be raised. New theories were put forward in the 1960s, principally by J. D. Hewlett, also in the USA. As outlined in Chapter 3, these theories place greater emphasis on interflow, and modern hydrograph analysis tends to recognise two components – a **baseflow** component, consisting usually of groundwater flow and slow interflow (plus, in some cases, meltwater and water from surface storage) and a **quickflow** component, derived from rapid interflow, any surface runoff and any rain that falls directly on the river channels. The division tends, therefore, to be based on the length of time the various components take to reach the main drainage channels, rather than on their route. As will be seen shortly, this is not always satisfactory to the hydrogeologist.

The hydrograph segment in Figure 8.9 commences (time A) at the end of a long period without rain, when we can assume that all the flow in the stream was baseflow, derived largely from groundwater storage. As this groundwater discharge took place, the amount of groundwater stored in the aquifers decreased, leading to a fall in the potentiometric surfaces and a reduction in the hydraulic gradient and hence in the flow of groundwater to the river; as a result the hydrograph (GH) shows a gradual decrease in flow.

Between B and C a period of rainfall occurred. Some of the rainfall was disposed of as quickflow, reaching the surface water-courses rapidly and leading to an increase in river discharge to a peak value at J. This increased flow then began to diminish, and it can be assumed that by time D the flow was once again all baseflow. The discharge is now greater than it was before the rainfall, however, indicating that there has been an increase in the amount of water stored in the aquifers; some of the rainfall has therefore infiltrated into the aquifers.

The suggested division of the total discharge into baseflow and quickflow is shown by the dashed line in Figure 8.9. This dividing line is termed the **baseflow-separation curve**; it could be drawn in a variety of ways and according to a variety of rules, the final choice invariably being subjective. The possibility exists nowadays of drawing it by computer; computer methods achieve consistency, but they still have subjectivity built into the programs they use, because the programs have to be written by a human.

The discharge hydrograph is a graph of discharge (volume of water per unit of time) plotted against time. It therefore follows that the area under the graph between any two times (e.g the area AGHJKLE in Figure 8.9) represents the total volume flowing past the gauging station during the time interval AE. Similarly, the area beneath the baseflow-separation curve (the area AGHXKLE, for example) represents the volume of baseflow passing the gauge in that time interval.

Hydrograph analysis thus provides a way of knowing how much groundwater is flowing from a catchment area upstream of the gauging station in a given time period – a year, for example. This is interesting, but much more significant is the fact that if groundwater is leaving the catchment it must have entered it – in other words, the volume of water leaving the catchment as baseflow must have entered as infiltration. Dividing this volume by the catchment area enables us to express the infiltration in millimetres, so that we can compare it with rainfall and evapotranspiration figures. Doing this calculation over one year can be misleading, as there may be changes in the amount of groundwater in storage in the aquifers, especially if there have been unusual rainfall conditions. It is therefore advisable to average the results over a period of at least five years and preferably more than ten. Used sensibly, this is the best general technique for assessing the infiltration over an area. However, it can work satisfactorily for the hydrogeologist only if the separated component represents infiltration that reaches the aquifers; this is why an arbitrary division into baseflow and quickflow may not be entirely satisfactory. It is also essential to know the area of the catchment contributing the baseflow; if the groundwater and surface-water catchments do not coincide this may be difficult to determine.

Figure 8.9 was a hydrograph for an imaginary catchment containing permeable and impermeable material. Figure 8.10 shows a real hydrograph for a Scottish stream draining a catchment underlain almost entirely by impermeable metamorphic rocks (mica-schists). The inability of rainfall to infiltrate to any depth, and the lack of groundwater storage, means that a rainfall event produces a rapid rise in stream flow, which is followed by an almost equally rapid decline at the end of the rainfall. In dry weather the flow almost ceases, being sustained largely by minor baseflow contributions from peat and superficial deposits. This type of river response is termed 'flashy'; in areas like this, flooding is likely after heavy rain, and catchments are characterised by numerous streams and tributaries.

In contrast, Figure 8.11 shows a hydrograph from a river draining a

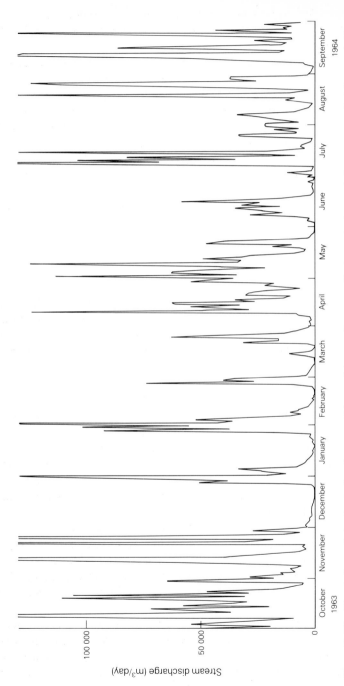

Figure 8.10 The drainage behaviour of an impermeable catchment A stream-discharge hydrograph for a stream draining a catchment area underlain by metamorphic rocks in the Western Highlands of Scotland. (Peak discharges are not shown above 130 000 m^3/day.) (Reproduced by permission of the Director, British Geological Survey, and based on data supplied by North-West Scotland Hydro-Electric Board.)

catchment underlain by Chalk. Here there is hardly any surface runoff or interflow, because almost all the rainfall infiltrates into the Chalk and emerges as baseflow.

In permeable catchments like this one there is little in the way of surface drainage; virtually all streams are gaining streams whose valleys intersect the water table. Many factors other than geology influence the response of streams to rainfall, but aquifers exert a stabilising influence on streamflow, taking water into storage during periods of heavy rainfall (thereby reducing the possibility of flooding) and releasing it slowly during dry weather, thus maintaining streamflow which is available for water supply, dilution of sewage effluent, navigation, recreation (including fishing), and so on. Statistics that consider only the amounts pumped directly from aquifers can therefore seriously underestimate the importance of groundwater.

What happens in a drought?

The importance of groundwater in maintaining streamflow becomes more apparent in a drought. Having said that, it must be admitted that there is no agreed definition of the word 'drought', since dry weather affects different people in different ways. A dry period in the spring may please most people but could be disastrous to farmers; similarly the farmers may welcome a relatively dry winter which worries water engineers, who know that reservoirs and aquifers are not being fully replenished ready for the demands of next summer. For this reason meteorologists, agriculturalists and hydrologists all have their own definitions of drought. The general public tend to welcome dry weather until it persists to the point where it interferes with their lives – perhaps by leading to a shortage of fresh vegetables, or to a ban on watering of gardens, or (in Britain very exceptionally) to a rationing of water supplies to domestic consumers.

In 1975 and 1976 there occurred over most of Britain and large parts of Western Europe a pattern of weather that was accepted by all those who experienced it as a drought, and in parts of the affected area the most severe drought since records began. This consensus of opinion makes this drought, although not typical, a useful one to consider as an example.

The 1975–76 drought in Britain was not a single unusual event, but a combination of unusual events. The winter of 1974–75 was somewhat wetter than average, but from May to August 1975, England and Wales received only about two-thirds of the average rainfall for that period.

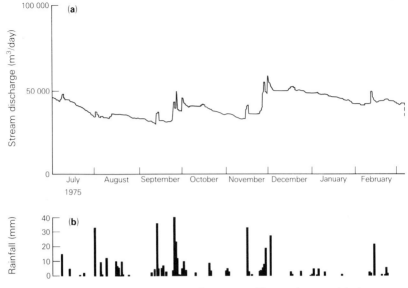

Figure 8.11   The drainage behaviour of a permeable catchment   (a) A stream-discharge hydrograph for a stream draining a catchment area in Hampshire underlain by Chalk. (b) Rainfall at a nearby rain gauge (amounts less than 1 mm have been omitted). (Based on data supplied by Southern Water Authority.)

(The 'averages' here refer to average precipitation during 1916–50, which is a standard period for this purpose.) Relatively dry weather continued throughout the winter of 1975–76 – one of the driest winters of the last century – with England and Wales receiving little more than 60 per cent of average precipitation. The hot dry summer of 1975 resulted in high rates of potential evapotranspiration, leading to the development of large soil-moisture deficits over most of Britain – deficits that were slow to diminish during the following dry winter.

By February 1976, soils in many areas had only just about returned to field capacity, meaning that little or no extra water was available as recharge to aquifers. This meant that in many areas groundwater storage and therefore groundwater levels (potentiometric surfaces) continued to decline during the winter of 1975–76, instead of showing the usual rise. Fortunately, the wet weather of the winter of 1974–75 had resulted in high groundwater levels in many aquifers at the beginning of 1974, so that despite the dry summer of 1975 and the lack of recharge during the succeeding winter, groundwater storage was not seriously depleted.

Spring and summer of 1976 continued to be dry (Fig. 8.11). The summer was also exceptionally hot and sunny, with record high temperatures during late June and early July and record sunshine levels for August at

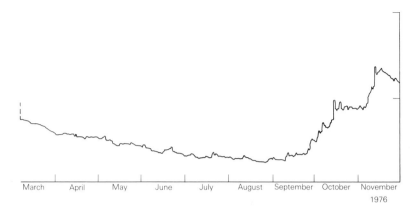

March	April	May	June	July	August	September	October	November
								1976

many observation stations. These conditions favoured high evapotranspiration, and soil-moisture deficits soon reached the point at which there was little or no moisture available for plant growth.

Groundwater levels continued to decline, so that springs and seepages either ceased to flow or flowed at a diminished rate; since all other flow had disappeared, this reduction in baseflow resulted in many streams and rivers almost ceasing to flow. Similarly, the lowering of water tables meant that many shallow wells became dry, in some cases for the first time in living memory. In general, however, deep wells such as those used for public supplies proved reliable. In contrast, some areas of the country dependent on rivers or surface-water reservoirs for their water supplies suffered shortages, with many reservoirs suffering from a combination of reduced recharge in the 1975–76 winter and high evaporative losses in the dry summer of 1976. In some of these areas water rationing was imposed.

One possible effect of droughts is that the decline of groundwater storage in unconfined aquifers can sometimes result in a water table being lowered below the bed of a river, so that what is normally a gaining stream becomes a losing stream. An apparent example of this occurred in the drought of 1975–76, resulting in the much-publicised Thames 'leak'. During the driest period of 1976, it was discovered that the flow of the Thames between Eynsham (near Oxford) and Dorchester, some 40 km downstream, was decreasing. A study indicated that much of the decrease could be accounted for by evaporation, but that the remainder appeared to be caused by water seeping from the river into the adjacent aquifer.

Whatever else the drought did, it drew attention to the fact that water is an important commodity which, even in a country like Britain with a temperate and humid climate, cannot always be taken for granted. With the general public being made uncomfortably aware of this fact, it was not long before politicians at all levels and of all political shades were competing with each other for news coverage as they either praised or criticised the water authorities for coping or not coping with the problem. The news media were able to fill in the 'silly season', reporting all these aspects.

But all things come to an end, and the drought ended with a September and October that were the wettest in England and Wales for over 250 years, and with a generally wet winter. Soil-moisture deficits disappeared, aquifers and reservoirs were replenished, and most of us went back to washing our cars and not worrying about water shortages.

Perennial drought – deserts

There are some areas of the world where drought is not an unusual occurrence, as it is in Western Europe, but is the normal state of affairs; these areas are the **deserts**. As in the case of droughts, there is no generally accepted definition of a desert. Regions with low precipitation (less than 250 mm per year) can be found at most latitudes, but the areas that most people would regard as deserts have low precipitation and lie between latitudes 15° and 50° north and south of the Equator.

These areas are characterised by their dryness, lack of vegetation and sparse population. Some are sandy, with or without dunes; some are covered with gravel, and others are predominantly bare rock. Surface drainage in true deserts is virtually non-existent, being present only after heavy rainstorms which may occur less than once in a hundred years; in semi-deserts, such as some of the intermontane regions of the United States, rainfall is more frequent and gives rise to ephemeral streams. Semi-arid areas occur in North Africa, on the edge of the Sahara, and here these ephemeral watercourses are termed **wadis**. Some recharge to aquifers beneath deserts may take place through the beds of temporary watercourses; elsewhere, the rainfall will usually be absorbed to satisfy the soil-moisture deficit before it can infiltrate far into the ground.

Many of the world's large arid and semi-arid areas are underlain by permeable materials, either deposited in intermontane basins and valleys as a result of erosion of the surrounding highlands (as in the case of many of the desert aquifers of the southwestern United States), or present as older, more widespread formations, often consolidated or partly so. Given adequate recharge, these formations can be valuable aquifers, vital to the development of these regions. In some areas recharge does take place, either as the result of infrequent but intense rainfall which can locally satisfy soil-moisture deficits or as a result of more regular rainfall on highlands bordering the desert. In some cases, studies of the isotopes (Ch. 11) present in the water indicate that it infiltrated long before the present day. In the Libyan Desert, for example, there are enormous reserves of water in Tertiary sands. It appears that much of this water infiltrated between about 35 000 and 15 000 years ago. It would take thousands of years for groundwater to percolate to the central Libyan Desert from the most likely present-day recharge area in the Tibesti Mountains to the south-west, so this could be one explanation for the age of the groundwater. There is considerable evidence however that infiltration occurred into the desert itself at periods in the past (including the period between 35 000 and 15 000 years ago) when the climate of the area was colder and wetter than at present. These **pluvial** (rainy) periods were related to the glaciations that affected Europe during the Ice Age.

Although limited recharge may be occurring at the present day in several deserts in the Middle East, most of the fresh water they contain probably infiltrated during pluvial periods.

Because this water entered the aquifers in the past, it is often described as 'fossil water'. Like other groundwater it can be abstracted and exploited, but unlike other groundwater it is not being replenished at the present time. Its exploitation is therefore analogous to that of any other non-renewable mineral resource, such as oil, coal or copper, and for this reason abstraction of 'fossil water' is referred to as 'groundwater mining'.

Groundwater quality in desert areas is frequently poorer than in more humid areas. There are a number of reasons for this: predominant among them is the fact that high evaporation rates tend to concentrate soluble salts at the surface, ready to be dissolved by infiltrating water resulting from occasional heavy precipitation. A second major factor is that the limited recharge results in water being 'in residence' in the aquifer for long periods, giving it ample time to dissolve any soluble material present.

However poor its quality may be, however brackish it may taste to those accustomed to the mains water supplies of Europe or North America, this groundwater is vital to desert dwellers. Where groundwater comes to the surface naturally in a desert, the resulting spring or

(a)

(b)

Figure 8.12 Kufra Oasis, Libya (a) Aerial view of the desert near Kufra. In the foreground is the vegetation of the Kufra Oasis, which exists as a result of ground-water discharging from the Nubian Sandstone aquifer. The dark circles in the distance are cultivated areas irrigated by sprinklers; each circle is centred on a borehole draw-ing water from the Nubian Sandstone, and is swept out by a centre pivot irrigator (a sprinkler bar which rotates around the borehole). (b) A centre pivot irrigator and wheat in one of the irrigated areas shown in Figure 8.12a. (Photographs by W. M. Edmunds.)

seepage usually leads to growth of vegetation and the formation of an **oasis** (Fig. 8.12a). Oases occur for a variety of reasons; sometimes wind has removed the desert sand down to the level of the water table, when the resulting seepage of water and consequent vegetation lead to the stabilising of the floor of the depression. Other cases arise where confined groundwater is brought to the surface along fault zones or where erosion has removed the overlying confining bed. The quality is not always poor; at the Kufra oasis in Libya, for example, the groundwater contains less dissolved material than does much of the groundwater supplied for public use in Britain.

'Mining' of groundwater from these arid regions obviously must be carried out with caution. If development of resources is too rapid, quality may be endangered and supplies may be exhausted with little benefit to the inhabitants. Developed carefully, the vast reserves of groundwater in some of these regions can form the basis for agricultural or industrial communities (Fig. 8.12) which may be able to continue for many decades – perhaps until the time when water can be brought to them economically from large ocean-desalination plants.

Selected references

Burdon, D. J. 1982. Hydrogeological conditions in the Middle East. *Q. J. Engng Geol.* **15**, 71–82.

Davis, S. N. and R. J. M. De Wiest 1966. *Hydrogeology*. New York: Wiley. (See especially Ch. 12.)

Doornkamp, J. C. and K. J. Gregory (eds), 1980. *Atlas of drought in Britain 1975–76*. London: Institute of British Geographers.

Hewlett, J. D. and A. R. Hibbert 1967. Factors affecting the response of small watersheds to precipitation in humid areas. In *Symposium on forest hydrology*, W. E. Sopper and H. W. Lull (eds). Oxford: Pergamon.

Horton, R. E. 1933. The role of infiltration in the hydrologic cycle. *Trans. Am. Geophys. Union* **14**, 446–60.

Ineson, J. and R. A. Downing 1964. The groundwater component of river discharge and its relationship to hydrogeology. *J. Instn Water Engrs* **18**, 519–41.

Ineson, J. and R. A. Downing 1965. Some hydrogeological factors in permeable catchment studies. *J. Instn Water Engrs* **19**, 59–80.

Rodda, J. C., R. A. Downing and F. M. Law 1976. *Systematic hydrology*. London: Newnes-Butterworths. (See especially Ch. 6.)

Ward, R. C. 1975. *Principles of hydrology*, 2nd edn. Maidenhead: McGraw-Hill. (See especially Ch. 8.)

Ward, R. C. 1984. On the response to precipitation of headwater streams in humid areas. *J. Hydrol.* **74**, 171–89.

Wright, E. P., A. C. Benfield, W. M. Edmunds and R. Kitching 1982. Hydrogeology of the Kufra and Sirte basins, eastern Libya. *Q. J. Engng Geol.* **15**, 83–103.

9   Water wells

We do not know where, when or how the first well was sunk. If you had been one of the earliest members of the human race, your water supply would probably have been a river or a spring. If, in some period of dry weather, the spring had gradually ceased to flow or the river had begun to dry up, you would not have been able to explain the facts in terms of a falling water table, but perhaps it would have occurred to you to dig a hollow near the spring; or, noticing that parts of the river bed remained damp, you might have been inspired to dig a hole to reach the water which you would have found just below the surface.

However it began, the digging of wells was an established fact by Old Testament times (Genesis 26). Many Middle Eastern wells were shallow, tapping groundwater a few metres down in wadis or depressions, but some went much deeper. Joseph's Well, near Cairo, dating from the 17th century BC, was sunk to a depth of 90 m in consolidated rock, with a pathway around the sides allowing donkeys to go more than halfway down.

Another ancient Middle Eastern arrangement for collecting ground-water was the 'qanat' or 'kanat'. Kanats appear to have originated in Persia about 3000 years ago; they were horizontal or gently sloping galleries, up to 30 km long, which intersected groundwater in alluvial-fan material on the sides of mountains and conveyed it, underground, to discharge on the arid plains. Vertical shafts at intervals provided access and ventilation for the workmen digging the tunnel. Kanats were con-structed in suitable areas throughout the Middle East, and are still in use today.

The Ancient Chinese developed techniques for drilling, rather than digging, and by sustaining a slow rate of progress for several years could drill boreholes to remarkable depths – some accounts claim as deep as 1500 m. These deep holes were drilled to obtain brine rather than fresh water.

In Europe, early wells were dug shafts. The discovery of overflowing or 'artesian' conditions in Artois and other areas of France and Belgium, and in England, led to the development of borehole-drilling methods to tap these deeper supplies; for a time the term 'artesian well' was used for any well that was bored or drilled, as opposed to those which were excavated.

I am going to follow a convention of referring to wells which are dug or excavated as **shafts** and those which are drilled as **boreholes**. Other distinctions are possible, based on diameter or purpose; all are sensible, but the one I have chosen is convenient for present purposes.

For many people in Britain the word 'well' probably conjures up a picture of a shaft, a metre or so in diameter, lined with stone blocks or bricks, and topped by some form of windlass by means of which a bucket can be lowered on a rope and hauled up again full of water. The well will be surrounded by a wall to prevent animals or children from falling into it, and will probably be of great antiquity and little present-day relevance. The term 'oil well' on the other hand would probably invoke an image of a hole drilled deep into the Earth by modern machinery of great complexity and expense; such is the erroneous relative status of the oil and water industries in the public mind! This arises not from innate ignorance on the part of the public, but because the news media present the oil industry as dynamic and even glamorous; they rarely present the water industry at all unless it is in trouble.

Pumps

However the well is constructed (and more of this in a while), water must be brought to the surface by artificial means unless it overflows naturally. The bucket and rope may have been adequate for a cottage supply, but most modern wells use pumps. A hundred years ago, most pumps in water wells operated on a reciprocating-piston principle. The upward stroke of a piston in a vertical cylinder drew water into the cylinder, where a simple valve arrangement prevented it from being expelled as the piston moved back down; the next upward stroke forced this water from the cylinder and up into the delivery pipe leading to the surface. Small versions of these pumps could be worked by hand, and they can still be seen in some English villages; such handpumps provide the basic water supply in the rural areas of some developing countries like Bangladesh. Larger piston pumps were powered by steam engines in Victorian English pumping stations, and a few still survive in operating condition.

With the advent of high-speed rotary power units such as electric motors, pumps operating on a rotary principle came into their own. The most important category of these pumps is the **centrifugal** type. In a centrifugal pump, a wheel with vanes – called an **impeller** – rotates causing water inside the pump to move around and outwards (Fig. 9.1a). This results in an increase of pressure at the outer wall of the pump and a decrease near the centre of the impeller. Water is thus drawn through the pump from the centre to the edge, where it leaves through a delivery

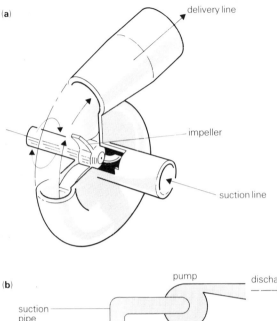

(a)

delivery line

impeller

suction line

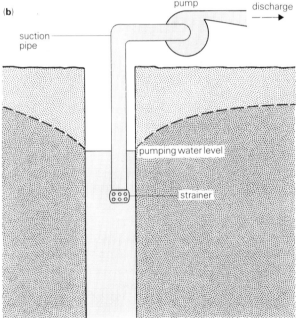

(b)

pump

discharge

suction pipe

pumping water level

strainer

Figure 9.1   Centrifugal pumps   (a) The principle of the centrifugal pump. (b) A centrifugal pump being used to pump water from a shaft. When pumping from a borehole, the suction pipe is often attached direct to the top of the borehole casing. In either case, the vertical distance between the pumping water level and the pump is limited to a maximum of about 7 m.

pipe. The pump is in effect transferring energy from the motor to the water.

Centrifugal pumps can be mounted at the surface near a well, with the inlet or suction pipe running down the well to below the water level. The operation of the pump reduces the pressure in the suction pipe (Fig. 9.1b) below atmospheric pressure. Because the atmosphere is exerting its full pressure on the surface of the water in the well, water is forced up the suction pipe. If the pump could reduce the pressure at its suction to zero – in other words, create a perfect vacuum – then the atmospheric pressure could raise the water to a height of about 10 m above the water level in the well. In practice, because of mechanical limitations and the possibility of pockets of water vapour (called 'cavities') forming in the water, these pumps cannot operate when the water level is more than about 7 m below the pump.

Where greater lifts are required, the problem is usually solved by putting the centrifugal pump under water with its impeller axis vertical. The pump can then be driven by a vertical shaft running down the well and turned by a motor at the surface; more commonly nowadays a waterproof electric motor is installed beneath the pump (where it is cooled by the water flowing past it) and drives it direct. This type of pump (Fig. 9.2) is called an **electric-submersible**. If the water has to be raised more than about 10 m from pumping water level to the surface, several impellers are used, one above the other, each imparting energy to the water. Each impeller is called a **stage**. Five to ten stages are most common, but twenty or more stages can be used for very large lifts. Because the impeller design for these pumps is usually of a type called a turbine, these deep-well pumps are often referred to indiscriminately as **turbine pumps**.

One other form of pumping that deserves mention is **air-lift pumping**. In this method (Fig. 9.3) air from a compressor is injected into a delivery main in the well and bubbles up through the water. The resulting frothy mixture of air and water is less dense than the water in the well, so the column of froth in the delivery main must increase in length if the pressure at its base is to be the same as the pressure of the water in the well. If the column of froth could become long enough, the pressures would be equal. If the system is correctly designed, however, the air–water mixture will overflow at the surface before this equilibration can take place; the pressure (and hence the static head) at the bottom of the delivery main will always remain less than that in the well, and water will flow from the well into the delivery main. This water in turn will be converted into froth and the process will continue. This discharge causes a drawdown in the well which in turn causes water to flow into the well from the aquifer.

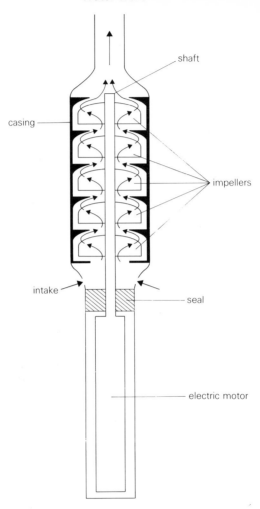

Figure 9.2   The principle of an electric submersible pump   The electric motor rotates the shaft which carries the impellers. Each impeller in effect operates as a centrifugal pump, drawing water in at its centre and accelerating it outwards and upwards. A seal prevents water from entering the motor.

Air-lift pumping is inefficient and therefore uneconomical for long-term use. It is ideal for pumping a well for a short period, perhaps immediately after its completion, or to save the expense of purchasing a suitably sized pump. It is sometimes used as a long-term method when the water is corrosive, or when it contains sand which would damage the moving parts of a conventional pump.

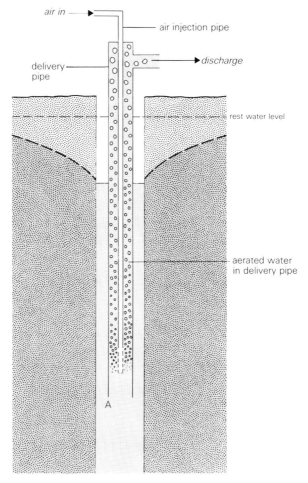

Figure 9.3 The principle of air-lift pumping The pressure at A is less inside the delivery pipe than outside, because of the low density of the aerated water. Water therefore flows into the delivery pipe from the well.

Well design

In consolidated rocks that will stand without support, a well can be very simple, and 'design' hardly enters into the matter. Unconsolidated sediments such as sand or gravel – which are often good aquifers – present a problem, as anyone who has tried to dig a hole on a beach will know; the sand, particularly where it is saturated with water, keeps collapsing into the hole. Above the water table in a well this problem can be overcome. Shafts can be lined with masonry, with pre-cast concrete rings, or with concrete poured in place. Boreholes are usually supported by inserting lengths of pipe, called **casing** or **lining tubes**, of diameter slightly smaller than the drilled diameter of the borehole. The space between the outside of the casing and the borehole wall is usually filled with a thin concrete called **cement grout.** In addition to providing support, this prevents dirty or polluted surface water or soil water from running into the well outside the casing after heavy rain; wells in consolidated rocks are usually given this **sanitary protection** for a few metres below the surface.

Below the water table the insertion of unperforated linings would support the surrounding aquifer but would also prevent the entry of water – not a desirable state of affairs! The well-sinkers who dug some of Britain's older wells got around this problem neatly, if laboriously, by constructing a lining of stones without mortar, like a dry-stone wall. The technique used in modern boreholes is to insert a lining tube that is perforated in such a way as to permit groundwater to flow from the aquifer into the well and at the same time to prevent aquifer particles such as sand grains from entering the well.

The choice of lining will depend on the aquifer. If the aquifer is consolidated but fractured, requiring support to prevent large blocks of rock from falling into the well, then casing with large circular perforations or crude slots will be adequate. If on the other hand the aquifer is loose sand, then a special lining called a **screen** is used. Screens have fine slots, and the appropriate slot size must be specified to suit the particular aquifer. Samples of the aquifer are passed through a series of sieves to determine the sizes of the sand particles, and the screen is chosen so that the slots are as large as possible (to permit the water to flow through the screen with minimum head loss) while being small enough to prevent the aquifer particles from moving into the well. Some of the finest particles will enter the well when it is first pumped – a process called **development** – but thereafter the well should produce sand-free water.

It sometimes happens that the aquifer is composed of particles which are so fine that to exclude them the screen slots would have to be impractically narrow. In such cases coarser slots are used, and an envelope of coarse sand or gravel, called a **filter pack,** is poured into the annular space between the aquifer and the screen. The filter pack retains the aquifer particles outside its outer boundary, and the screen slot size is chosen to prevent the filter pack itself from passing through the slots. A filter pack theoretically needs to be only a few grains thick to function effectively, but it is generally impossible to place sand or gravel in an annulus less than about 60 mm to 80 mm wide. Figure 9.4 shows a theoretical design for a well that draws water from three aquifers – a fine sand that requires a filter pack, a coarse sand, and a sandstone that does not need support. Development of the well causes removal of the finer particles from the coarse-sand aquifer, leading to the formation of a coarser zone – sometimes called a natural filter pack – immediately around the screen.

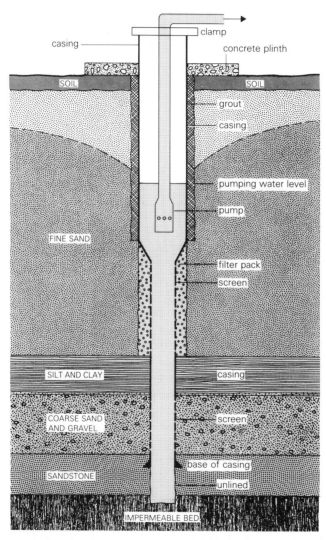

Figure 9.4 Well design A theoretical well completion in three aquifers: fine sand, requiring a screen and a filter pack; coarse sand and gravel, requiring a screen but no pack; and a consolidated sandstone, requiring no support.

The filter pack

There have been many studies of relationships between sands and filter packs, and numerous published recommendations regarding the ideal size of grains which should be present in a filter pack in order to support a given aquifer. These generally advise that the grains of the filter pack should be x times larger than the grains of the aquifer, where x is a number usually between 5 and 10. Some recommend that the filter pack should contain grains as nearly as possible of one size, while others advise that the pack should contain several grain sizes, i.e. be graded. Most of these recommendations tend to ignore the fact that what really matters is the relationship between the grain sizes of the aquifer and the void or pore sizes in the filter pack; the latter must prevent the aquifer particles from moving through the pack while permitting water to pass through with minimum head loss. So long as this is achieved, it matters very little whether the pack is uniform or graded.

Designing for maximum well efficiency

It sometimes surprises people to learn that wells in a particular area are several tens or even hundreds of metres deep, when the water table or potentiometric surface is only a few metres below ground level. There may be several reasons for this. In the case of a confined aquifer the well must obviously penetrate the full thickness of the confining layer to encounter water, even though the water may then rise to the ground surface. But even in an unconfined aquifer, it is not sufficient for a well just to reach water. If a borehole is drilled just to the water table, then the moment it is pumped – assuming that there is sufficient depth of water for the pump to operate – the water table will be lowered and the well will go dry. So the well must have an allowance for drawdown, which is why in Figure 9.4 the top of the screen is a long way below the water table – there is no point in putting expensive screen and filter pack in the cone of depression. The well need not penetrate the full thickness of the aquifer, but the well designer cannot expect to have the benefit of the total transmissivity of the aquifer if the well penetrates only a fraction of the aquifer thickness.

A related point is that the well must provide an adequate intake area – the area through which water flows into the well. This open area depends on the diameter of the well in addition to the thickness of aquifer it penetrates. A common statement is that doubling the diameter of a well will decrease the drawdown by only about ten per cent for the same yield – in other words, it is a waste of money to drill a large-diameter well. The statement is correct as far as the aquifer-loss component of the drawdown is concerned, but can be misleading when the well-loss component is also considered. The importance of intake area can be visualised most clearly in the case of a well completed with a screen; if water is flowing into the well at Q m^3/s and the total area of the screen slots or perforations is A m^2, then the entrance speed v of the water through the openings is given by

$$v = Q/A$$

Experience suggests that v should not exceed 0.03 m/s (although some experiments have suggested that this value is too conservative), or too much energy will be wasted as a result of turbulence as the water enters the well and turns to flow towards the pump. Given this limitation the designer, knowing the discharge Q that is required from the well, can calculate the area of slots needed. This area depends on three factors: the length l and the radius r_w of the screened interval control the total surface area of the cylindrical lining ($2\pi r_w l$); the open area depends on what percentage of the surface area of the screen consists of slots. The most efficient screens have as much as 50 per cent open area; these screens are expensive but may more than justify their extra unit cost by allowing the designer to use a shorter length of screen or one of smaller diameter.

In consolidated rocks, for which linings are not required, the need for adequate intake area to reduce well losses still exists but the problem is usually less than in rocks for which screens *are* required. One aspect in which diameter is important in all aquifers, however, is in accommodating the pump. The upper part of the well, which serves to accommodate the pump (Fig. 9.4), is frequently of larger diameter than the lower or intake portion. Because most of the upper section is within the cone of depression, it is lined with plain casing, which is cheaper than perforated casing or screen; it is kept as shallow as possible, to reduce the expense of drilling and lining this large-diameter hole. As submersible pumps become slimmer, this expense can be reduced further.

Drilling methods

Designing a well is one thing; drilling and completing it to that design is another. Not the least of the problems is the conflict between theory and practice – the theoretical knowledge of the hydrogeologist or engineer who knows what he wants, and the practical knowledge of the driller who knows what is possible with the equipment available.

If you want to make a hole in a wall you have a choice of two methods: you can use a rotary bit in an electric drill, or you can use a hammer and chisel. The first is quick, but requires relatively expensive and sophisticated equipment; the second uses equipment that is simple and cheap, but it tends to be time-consuming. The same essential techniques, with the same basic advantages and disadvantages, are available for drilling boreholes in the **rotary** and **percussion** methods.

In the percussion method a heavy drill bit called a **chisel** on the end of a wire rope is alternately raised and dropped to break or disaggregate the rock. The fragments are formed into a slurry either with groundwater or, in unsaturated or impermeable material, by adding a small amount of water to the hole. The slurry is periodically removed using a **bailer**, which is in essence a length of steel pipe with a flap valve at its lower end. As the hole is deepened more cable is fed out from a drum, so that the chisel just strikes the bottom of the hole with each impact.

In its simplest form the technique uses a **drilling string** (the assembly of drilling tools which is lowered into the hole) consisting of little more than the chisel and a rope socket which attaches it to the end of the cable. The percussive or 'spudding' action is provided by the driller alternately engaging and disengaging the clutch on the cable drum, so that the bit is lifted about a half to one metre and then dropped again. This simple technique is widely used for drilling site-investigation boreholes.

The larger percussion rigs used to drill water wells have the spudding action provided by an oscillating beam called the **walking beam**, which is pivoted at one end and which has a sheave at its free end beneath which the cable passes (Fig. 9.5). The rocking motion of this beam alternately raises and drops the drilling bit. The bailer is carried on an additional line called the **sand line.** The drilling string comprises the bit or chisel; a sinker bar (in North America this is called a drillstem) to add weight and length and so help to drill a straight and vertical hole; a sliding link called the 'jars' to help in freeing the bit if it gets stuck; and the rope socket. The whole assembly may vary in diameter from 100 mm to a metre or more, and can weigh well over a tonne. Progress is usually good in consolidated rocks but in loose sand the borehole walls must be supported

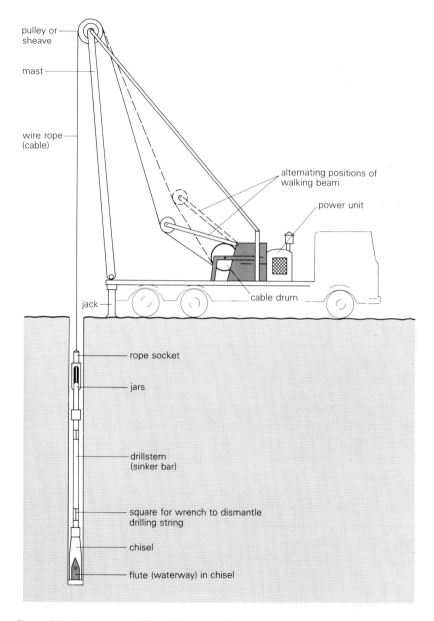

pulley or
sheave

mast

wire rope
(cable)

alternating positions of
walking beam

power unit

jack

cable drum

rope socket

jars

drillstem
(sinker bar)

square for wrench to dismantle
drilling string

chisel

flute (waterway) in chisel

Figure 9.5 Percussion drilling The essential parts of a truck-mounted cable-tool (percussion) drilling rig. The bailer and sand line are omitted for clarity.

by temporary casing, which is withdrawn once the permanent casing or screen has been installed; these operations can slow down the average rate of progress.

The drilling industry's version of the handyman's electric drill is the rotary drilling rig (Fig. 9.6). The bit on this machine rotates at the lower end of a hollow steel tube, which is composed of lengths of **drill pipe** screwed together. **Drilling fluid** – usually a mixture of water and clay known as 'drilling mud' – is continuously pumped down the drill pipe and through the bit. Its function is to cool and lubricate the bit, and to remove the rock fragments (cuttings) produced as the hole is deepened, by carrying them to the surface up the annular space between the drill pipe and the borehole wall. The pressure of the fluid also supports the borehole wall in unconsolidated material so that temporary casing is only rarely needed. In permeable rocks the liquid part of the drilling mud seeps from the borehole into the formation, leaving a **filter cake** of mud solids on the borehole wall which helps to prevent further fluid loss. At the surface the mud flows into a settling tank or pit where the cuttings are deposited before the mud is pumped back down the drill pipe.

The rotary motion is applied to the drilling string at the **rotary table**, which is a rotating bush with a square hole at its centre, turned by the rig engine. The top length of the drilling string is a special length of drill pipe, of square cross-section, called the **kelly**. This fits in the rotary table and is rotated by it. The weight of the drilling string is partly taken by a wire rope from the rig draw works (in essence, a powerful winch), and the kelly is allowed to slide down through the rotary table as the hole is deepened. When the full length of the kelly has slid through the rotary table in this way, the drilling string is pulled up by an amount equal to one length of drill pipe and the kelly is unscrewed. An extra length of drill pipe is then added before the kelly is reconnected, the drilling string being held meanwhile by steel wedges called 'slips' which fit into the rotary table. The drilling fluid is pumped down the kelly through a hose and a swivel arrangement which permits the mud to enter the kelly while the kelly is rotating.

Rotary rigs range in size and complexity from units capable of drilling holes about 100 m deep and which are usually mounted on tractors or lorry trailers, to the giant machines with masts about 60 m or 70 m high which are used for drilling oil and gas wells. Tall masts permit the use of long individual lengths of drill pipe, thus reducing the time spent in adding or removing lengths. They similarly speed up the process of installing casing. Any rig, whether rotary or percussion, must have the lifting ability to handle the long and heavy total lengths of drill pipe or casing which may have to be lowered into, or pulled from, the well at

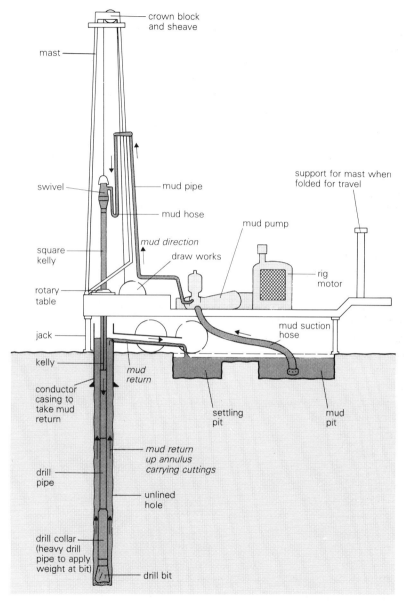

crown block
and sheave

mast

support for mast when
folded for travel

swivel

mud pipe

mud hose

mud pump

mud direction

square
kelly

draw works

rotary
table

rig
motor

jack

mud suction
hose

kelly

conductor
casing to
take mud
return

*mud
return*

settling
pit

mud
pit

*mud return
up annulus
carrying cuttings*

drill
pipe

*unlined
hole*

drill collar
(heavy drill
pipe to apply
weight at bit)

drill bit

Figure 9.6 Rotary drilling The main features of a trailer-mounted rotary drilling rig.

total depth; this lifting ability is usually what decides the depth of well that can be drilled by a particular rig.

Variations on the basic rotary method include the top-drive rig, in which the rotary power is transmitted directly to the top of the drill pipe by a hydraulic turbine which is lowered down the mast by the draw works as drilling progresses. Another variation uses reverse circulation, in which the drilling fluid (which in this case is usually water) is pumped down the annulus and returns up the drill pipe, which is of large diameter. This method is especially suitable for drilling large-diameter wells in gravels, as quite large cobbles can be brought up the drill pipe without being broken up.

Some rotary rigs use compressed air as the drilling fluid. Another use of compressed air is the method known as **downhole hammer** drilling. For this a small rotary rig is used, but the drilling action is essentially percussive. The drill pipe carries a tool rather like a roadworker's pneumatic drill: compressed air pumped down the pipe operates the cutting tool and carries the cuttings to the surface. The technique is especially suited to hard formations such as basalts; like the air-flush rotary method, its disadvantage is that most compressors are unable to provide air at sufficient pressure to operate the downhole tool and overcome a large head of water. It is therefore not possible to drill to any great depth below the water table with this method, unless the permeability is so low that the air-lift action of the returning air is able to keep the hole effectively pumped dry.

A relatively recent innovation is the downhole motor. In this drilling method the drill pipe remains stationary, serving merely to convey drilling fluid to a turbine which sits at the bottom of the drilling string, coupled directly to the bit. This method is gaining popularity in oil-well drilling, particularly in the USSR, largely because it makes possible the drilling of non-vertical holes with precise control over the hole orientation. It has not yet been applied to the drilling of conventional water wells, but it is likely that this application will come.

All of the drilling methods mentioned above have advantages and disadvantages, and no single method is superior to the others for all applications. The percussion method uses simple equipment, requires little water, and makes good progress in most formations except loose sands (which often 'run' into the hole faster than they can be removed by the bailer). A disadvantage is that in unconsolidated formations temporary casing must be installed and subsequently removed; in general, the rate of penetration is slower than that of rotary rigs in similar circumstances.

Conventional rotary drilling is suitable for most formations except those containing large pebbles or nodules in a loose matrix; the pebbles rotate with the bit instead of breaking up, and may deflect the drilling string away from the vertical. Formations containing large open fissures cause problems with **lost circulation**; the drilling fluid flows into the fissures instead of returning to the surface, and fibrous material must be added to the fluid to bridge the fissures. Rotary drilling is fast and the presence of the drilling mud removes the need for temporary casing, but the equipment is more expensive and more complicated, the filter cake may subsequently inhibit the flow of water into the well, and large quantities of water may be needed for drilling.

In some parts of the world shafts are still excavated. Their construction generally requires that people work inside the shaft, which can be dangerous unless adequate – and usually expensive – safety precautions are taken, such as shoring-up the sides to prevent collapse. These precautions and the labour-intensive nature of the work add greatly to the cost, so the method is rarely used except in countries where labour is cheap and safety standards are, unhappily, low. A major disadvantage is that digging usually cannot continue far below the water table, so construction must be undertaken at a time when the water table is at its lowest level – otherwise there is a continual likelihood of the well going dry. Shafts are usually dug in relatively impermeable material, where their large diameters make for useful amounts of storage; unfortunately, the large diameters may also allow their contents to be easily contaminated.

Sampling and coring

So far we have talked about drilling boreholes in terms of obtaining water, with the emphasis on the hole in the ground as the end product. Sometimes, particularly in the exploration stage of a project, it is the ground that is of interest rather than the hole. These are the times when samples of the aquifer and possibly of any confining beds must be obtained – perhaps for sieve analysis so that correct screens can be ordered or filter packs prepared, or perhaps for laboratory determination of permeability or some other property.

Any drilling method results in fragments of rock being brought to the surface, but the condition and value of the fragments varies with the method. The bailer of a percussion rig brings to the surface a sample whose depth of origin is usually known to within about a metre. The drilling fluid of a rotary rig takes a finite time to bring the cuttings to the surface – time that increases as the hole gets deeper and which may sometimes be over an hour. In that time the hole may have been deepened by 10 m or more, so that the cuttings arriving at the surface are from some way above the current bottom of the hole; allowance has to be made for this, and the cuttings have to be interpreted with caution. Drilling bits are designed to break up and grind away the rock as efficiently and as quickly as possible, so the fragments are often very small; the dust that returns to the surface from a downhole hammer, in particular, is virtually useless to the geologist.

If a relatively undisturbed sample from a precisely known depth is required, the method called rotary **coring** is usually employed. Nowadays

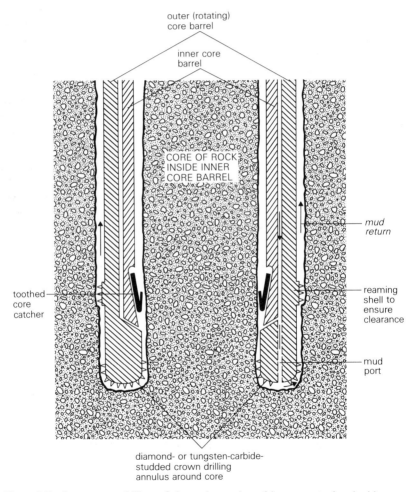

Figure 9.7 Rotary core drilling Schematic section of lower part of a double-core barrel. When the barrel is full, the core is broken off by increasing the speed of rotation and pulling up slightly.

this technique makes use of a double-tube **core barrel** (Fig. 9.7), which is run into the hole on drill pipe. The outer tube, which rotates, carries a diamond or tungsten-carbide cutting edge, which cuts an annular space around a central core of rock which is typically between 50 mm and 150 mm in diameter. As drilling progresses, the inner tube of the core barrel slides down over the core. The inner tube remains stationary and protects the core from the drilling fluid, which passes between the two tubes and out through special ports in the cutting edge or **crown**. Core barrels on the small rigs used for drilling water wells are usually about 3 m long; when this distance has been drilled and the barrel is filled, the barrel is withdrawn from the hole, the core being retained by a toothed **core catcher** at the lower end of the inner tube. The core is extruded from the barrel at the surface.

The need to withdraw the barrel and all the drill pipe each time the length of the barrel has been drilled means that core drilling takes much longer than conventional rotary drilling. It is therefore more expensive, so cores are usually taken only in exploratory holes, and then only at selected intervals. The method cannot be used in unconsolidated or very broken strata.

Primitive forms of core drilling, in which the core barrel was little more than a tube with a cutting edge, have been used for centuries. In Victorian times many wells were drilled in English Permo-Triassic sandstones using these simple core barrels; steel balls known as 'chilled steel shot' were dropped into the boreholes to aid the cutting action of the barrel. The resulting cores, which were sometimes a metre in diameter, were often dumped outside the pumping stations and can still be seen at some. Nowadays cores are carefully examined and are usually retained for special tests, including the permeability tests described in the next chapter.

Selected references

Davis, S. N. and R. J. M. De Wiest 1966. *Hydrogeology.* New York: Wiley. (See especially Ch. 8.)

Todd, D. K. 1980. *Groundwater hydrology,* 2nd edn. New York: Wiley. (See especially Ch.5.)

Vallentine, H. R. 1967. *Water in the service of man.* London: Penguin. (See especially Ch. 10.)

10 Measurements and models

Groundwater is part of the hydrological cycle, but the underground part of that cycle is constrained by geological controls. To study the physical hydrogeology or quantitative groundwater resources of an area three sets of factors have to be considered, requiring three different but inter-dependent fields of study: **geological**, to investigate the framework in which the groundwater occurs; **hydrological**, to investigate the input and output of water to and from that framework; and **hydraulic**, to investigate the way in which the framework constrains the water movement. None of these three is more important than the others; some hydrogeological investigations have suffered because one aspect was neglected.

Hydrological measurements

From the viewpoint of groundwater development, hydrological measurements are needed so that the water balance of the study area can be determined. In its simplest form the water balance can be expressed as:

water entering the area = water leaving the area ± any change in storage.

In more detail this is:

$$\left(\begin{array}{l}\text{precipitation} + \text{streamflow in} + \\ \text{interflow in} + \text{groundwater flow in} + \\ \text{artificial inflow}\end{array}\right) =$$

$$\left(\begin{array}{l}\text{evapotranspiration} + \text{streamflow out} + \\ \text{interflow out} + \text{groundwater flow out} + \\ \text{artificial abstraction of water}\end{array}\right)$$

$$\pm \text{ (any change in storage)}$$

If a long time period is chosen, beginning and ending at the same time of the year, it is usually safe to assume that storage changes can be disregarded. It is usual to take for the study area a river catchment, so there is no inflow of surface water or interflow from outside the area, although there may be inflow of groundwater if the surface-water and groundwater divides do not coincide (Fig. 8.2). Then the water balance becomes:

$$\left(\begin{array}{l}\text{precipitation} \\ +\ \text{groundwater flow in} \\ +\ \text{artificial inflow}\end{array}\right) = \left(\begin{array}{l}\text{evapotranspiration} \\ +\ \text{streamflow} \\ +\ \text{groundwater flow out} \\ +\ \text{artificial abstraction}\end{array}\right)$$

Precipitation can be measured using raingauges, and an average value calculated for the catchment. Evapotranspiration can be calculated from formulae if sufficient climatological information is available, or more direct estimates can be made using lysimeters. A **lysimeter** is an enclosed volume of soil, covered with natural vegetation and maintained under natural conditions, from which the amount of water being evaporated can be measured directly, usually by weighing the isolated soil and vegetation.

Streamflow, including the groundwater and interflow components, can be measured at a gauging station at the point where the river leaves the catchment. Groundwater may leave a catchment as percolation through an aquifer beneath the river; this is sometimes termed **underflow**. If sufficient information is available about the permeability, geometry, and hydraulic gradient, the underflow can be estimated using Darcy's law. Comparison of infiltration estimated from hydrograph analysis (Ch. 8) with estimates based on the difference between evapotranspiration and precipitation will usually indicate whether or not groundwater is entering from another catchment.

The remaining unknown, net artificial abstraction, can be calculated by comparing the withdrawals made at each individual well or river intake with amounts returned to the ground or drainage system, for example at sewage outfalls.

Working out the water balance for a catchment is in some ways like checking the financial balance of a household or any other economic unit – it helps us to see where the 'expenditure' is going. If having worked out each item in the equation independently, we put them together and the equation really *does* balance, it gives us some confidence that we know what is happening; if it fails to balance, then we may at least be able to see where more accurate measurements are needed.

Just as checking the household budget enables us to see whether we can afford to spend more or whether we should restrict expenditure, so consideration of the water balance enables us to see whether more use can be made of the catchment's water resources. If there is ample infiltration and ample baseflow to sustain the minimum flow required in the river, for example, then more wells can be drilled to draw on groundwater. It used to be argued that the maximum withdrawal of groundwater from a catchment must not exceed the infiltration into that catchment, so as not to cause a permanent lowering of the potentiometric surface, but arguments of that kind are not readily accepted now; lowering the water table may help to increase infiltration and may reduce unnecessary winter baseflow to streams.

Sometimes the long-term approach of ignoring changes in storage may be inadequate. If recharge is markedly seasonal and demand for water is greatest in the dry season, then the amount of storage in the catchment becomes of importance. For example, in parts of Asia, where recharge is restricted to the monsoon period and where water is needed for irrigation in the dry season, it may be necessary to compare the precipitation and evapotranspiration, and to calculate soil-moisture deficits and water-table changes, on a monthly or even a daily basis.

Geological measurements

An understanding of the basic geology of an area is vital to a study of its hydrogeology. Several tools are available to help the hydrogeologist achieve that understanding.

The most important of these tools is the geological map, which shows the distribution of various rock formations over an area and permits some understanding of the subsurface structure. Geological maps are available for all but the most remote parts of the world, though the detail and reliability are obviously variable. **Hydrogeological maps**, which combine information on basic geology with data on the hydraulic behaviour of the rocks and their usefulness for water supply, are now available for many countries. In Great Britain, hydrogeological and geological maps are produced by the British Geological Survey (formerly the Institute of Geological Sciences).

Aerial photographs are used in the preparation of maps, and they can be of direct use to the hydrogeologist. For example, where permeable rocks overlie impermeable material there are often springs or seepages (Fig. 8.3b); in arid regions in particular, these often show up because of the changes in vegetation that they cause. Seepages along faults and other major fractures or in other discharge areas (Fig. 8.3a) can often be detected from the air more easily than from the ground, and the presence of rocks with different permeabilities can influence the surface drainage in a way that is easy to see from the air. Photographs from satellites can now supplement aerial photographs, and films with special emulsions are available to enhance differences in soils and vegetation.

Photography from the air or from space is one of a group of techniques called **remote sensing** which make use of electromagnetic radiation to obtain information about parts of the Earth's surface and its near-surface structure. Usually these techniques produce an image of the ground surface; when the electromagnetic radiation being used is visible light, the image is of course an ordinary photograph. Imagery that uses some wavelengths of infra-red radiation is particularly useful in hydrogeological studies; in this method, called **line scanning**, equipment mounted in an aircraft builds up an image of the ground that depends on the amount of infra-red energy which the ground surface is reflecting or emitting. By making the survey at a time when solar heating effects are unimportant (usually just before dawn) it is possible to produce an image which depends on the temperature of the ground surface. Because of water's high specific heat (Chapter 4), wet areas tend to stay at a constant temperature while dry rock and soil heat up or cool down more rapidly; this means that seepages can be identified on the infra-red image. Groundwater tends to stay at the same temperature throughout the year, whereas surface water is warmer in summer and cooler in winter; infra-red imagery can use these differences to detect springs discharging into rivers or into the sea.

Mapping and remote sensing provide information about the ground surface. Some idea of subsurface conditions may be inferred from this information, but the only way to find out for certain what is below the ground surface is to drill a borehole. A single borehole provides information at only one place; as a compromise between drilling a network of expensive boreholes and relying on mapping and surface imagery, techniques of **geophysical exploration** are often used. These involve measurement of various physical properties of the ground and are widely used in exploration for metallic ores and for oil. The properties that can be measured include electrical resistivity, local variations in the intensity of the Earth's magnetic field, changes in gravitational acceleration (which is influenced by the thickness and density of the rock formations beneath the point of measurement), and variations in the speed with which sound waves are transmitted through rock layers. Study of the behaviour of sound waves forms the basis of the seismic methods. Many remote-sensing methods are airborne applications of geophysical exploration.

Although many geophysical techniques may be used (cost permitting) to study geological structures during hydrogeological studies, the technique most often used to study the distribution of groundwater is

resistivity surveying. Most common minerals are poor conductors of electricity, so the ability of a rock layer to conduct electricity depends almost entirely on the amount and conductivity of the water it contains. Measurements made between simple electrodes pushed into the ground can be used to determine the depth to the water table, provided that the geology is straightforward.

For most geophysical surveys, at least one borehole is needed so that the physical measurements can be related to the geology. Geophysical measurements can also be made in boreholes (Fig. 10.1); they are then generally known as **well logging** methods. Well logging involves lowering an instrument probe (a **sonde**) into the borehole and making measurements of physical properties of the surrounding rocks or of the borehole itself; a graph (called a **log**) of the property's variation

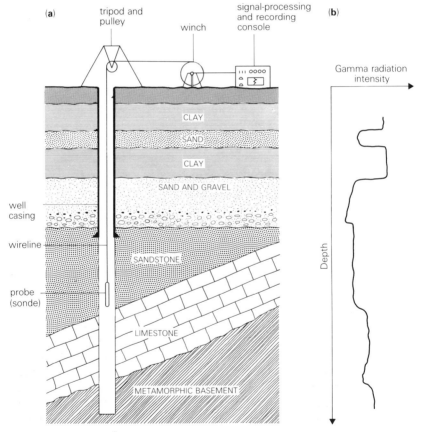

Figure 10.1 Geophysical logging   (a) The principles of geophysical logging. (b) A hypothetical gamma log from the strata at (a).

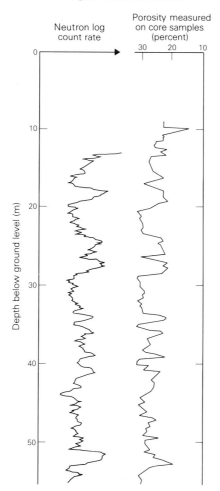

Figure 10.2 Porosity measurement Comparison of a neutron log (a) with laboratory measurements of porosity (b), made on core samples, from a borehole in Permian sandstone. (Reproduced by permission of the Director, British Geological Survey, neutron log by BPB Industries Ltd.)

with depth is produced. Properties measured in this way may include electrical resistivity, sonic velocity, and various radioactive properties. Measurements of the natural gamma radiation emitted by the rocks around the borehole are useful in correlating strata between boreholes because this radiation depends only on the lithology of the rocks; most other logging methods measure a response that depends on the porosity and on the properties of the fluid filling the pores. They are therefore useful to the oil industry (which developed most of them) and to hydrogeologists. Figure 10.2 shows a neutron log from a borehole in Permian sandstone and its similarity to a graph of porosity variation with depth; the neutron log responds to the presence of hydrogen nuclei, which are present in the water-filled pores. Porosity measurements can be made only when rock core or cuttings are available, whereas logs can be run in existing boreholes for which samples may no longer be available.

Some logging methods cannot be used in boreholes which are lined, and others will work only with certain types of fluid (e.g. drilling mud) present in the borehole. The methods discussed above measure properties of the rock in a zone extending, usually, a few centimetres or tens of centimetres from the borehole, and are referred to as **formation-logging methods**. Other logs can be run to determine the diameter of the borehole, and others are available to measure specific properties of the borehole fluid or its speed of movement. These methods however really belong under the heading of hydraulic measurements, discussed below.

Hydraulic measurements

When the hydrogeologist is satisfied that he has a reasonable understanding of the geology of an area, he can begin to concentrate on the hydraulic properties of the rocks. He will probably have gained some qualitative knowledge of these properties while studying the geology, but to develop the water resources effectively, quantitative measurements are needed. These measurements often proceed at the same time as water-balance studies, as the two are interdependent.

Aquifer properties
The hydraulic properties of interest (often called the **aquifer properties**) are the porosity of the aquifer or aquifers, which controls how much water is in storage; the storage coefficient, which controls how much of that water can be removed; the transmissivity (or hydraulic conductivity and effective thickness), which governs how readily that water can move through the permeable formations to wells and natural outlets; and the presence and position of **hydraulic boundaries**. Boundaries are in effect limits to the aquifer, because they prevent a cone of depression expanding beyond them. They do this either because they constitute an impermeable barrier (for example, where an aquifer meets impermeable material at a geological fault, as in Fig. 10.3a) or because they provide a source of effectively unlimited recharge. An example of a recharge boundary is a river in hydraulic connection with an aquifer: if a cone of

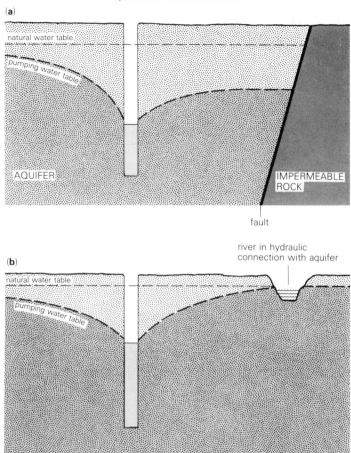

Figure 10.3 Hydraulic boundaries (a) A fault acting as a barrier boundary. (b) A river acting as a recharge boundary.

depression expands as far as the river, the river will prevent the cone expanding any further by supplying water to the aquifer (Fig 10.3b). In this way a gaining stream may locally become a losing stream. Both types of boundary cause the cone of depression to become markedly asymmetrical (Fig. 10.3).

Other variables that may have to be measured include the hydraulic conductivity of semi-permeable confining beds, through which water may 'leak' to or from the aquifer, and natural hydraulic gradients. These factors are often intimately related to the water-balance studies.

Some of the variables, such as hydraulic gradients and confined storage coefficient, can be measured only 'in the field', i.e. by in-place measurements in the ground. For other variables, such as porosity, permeability, and specific yield, a choice of laboratory or field measurements may be available. If laboratory methods are to be used, samples of the aquifer must be collected and returned to the laboratory in an unaltered condition. If the aquifer is unconsolidated this is usually impossible. Fissured aquifers also present a problem, since it is rarely possible to collect a representative piece of aquifer containing a fissure with its natural opening preserved. Usually, therefore, it is possible to make laboratory measurements only of the intergranular properties of consolidated materials.

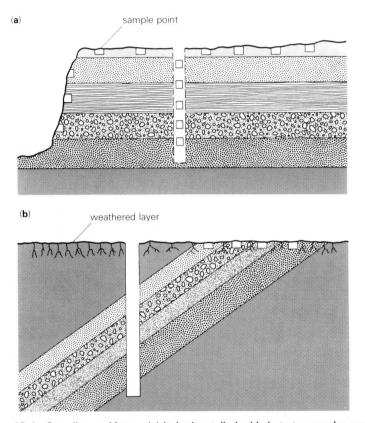

Figure 10.4 Sampling problems   (a) In horizontally bedded strata, samples can be collected from the whole sequence only from boreholes or from cliffs or quarry faces. (b) Where strata are inclined it may be possible to collect samples of the entire aquifer thickness from the outcrop. However, the outcrop material may be weathered and therefore unrepresentative of the aquifer at depth.

Sampling
The selection of samples requires careful thought. In general, the more samples the better, but a balance must be struck between spending large amounts of time and money studying too many samples and obtaining unreliable answers by studying too few. Samples must be representative of the bulk of the aquifer, and it is important not to allow bias to creep in when selecting them. If the aquifer is horizontal, then samples representing its full thickness can be collected only from boreholes or cliff sections (Fig. 10.4a); inclined formations offer the possibility of sampling the full thickness at the surface (Fig. 10.4b) but surface samples are usually affected by weathering which may have altered the properties to be measured. Borehole cores, if carefully sampled and handled, are therefore preferable.

Laboratory measurements
Laboratory measurements of porosity can be made in several ways. A simple way which is accurate enough for most purposes is to dry the sample, weigh it, and then evacuate it in a sealed container. A liquid of known density is then introduced into the container so that it covers and saturates the sample. The sample is weighed submerged in the liquid and, still saturated, in air; application of Archimedes' principle allows the bulk volume to be determined. The difference between the dry and saturated weights in air, divided by the specific weight (ϱg) of the liquid, gives the pore volume; the porosity can therefore be calculated. In practice, many samples can be evacuated and saturated simultaneously. By achieving controlled artificial drainage of the saturated sample, in a centrifuge for example, an approximate specific yield can also be determined.

Laboratory instruments for measuring permeability are called **permeameters**. In essence a permeameter is simply a device for holding a sample so that a fluid can be passed through it and Darcy's law applied to determine the hydraulic conductivity. Holding a sample of rock so that the fluid passes only through the rock and not through crevices between the rock and sample holder is difficult, and much research has gone into ways of achieving it. When dealing with consolidated aquifers most laboratories cut small cores, typically 25 mm to 50 mm in diameter and length, or cubes of similar size, from the borehole core or outcrop sample. At least two cores are usually taken from each sample, one with its axis parallel to the bedding and one at right angles to the bedding, so that permeabilities can be determined in these two directions. Cubes have the advantage that permeability can be measured in three directions on the same piece of rock by mounting the cube in the permeameter in different orientations.

Ideally, the fluid passed through the sample should be groundwater from that formation, and no other fluid should come into contact with the samples from the time they are part of the aquifer until the time the tests are finished. This is because groundwater is not pure water: it contains dissolved materials in various proportions and concentrations. The groundwater and the aquifer will be in approximate chemical equilibrium; if water with different dissolved material is introduced into the rock, the minerals in the rock may change. This is most likely to happen when clay minerals (present in many sedimentary rocks) are involved; the clay particles may swell or shrink as the water chemistry is changed, thereby altering the dimensions of the pores and hence the rock properties. (Even the action of drying samples for a porosity test can cause large changes, and sophisticated drying techniques may have to be used.) Values of permeability differing by a factor of a hundred or more can be measured on the same rock sample using different test fluids; if natural formation water (or a solution with the same dissolved constituents) is used for all measurements and preparation, the possibility of error is greatly reduced.

When dealing with 'clean' rocks (those containing negligible amounts of clay) the choice of test fluid is less critical, and distilled water or even a gas can be used for permeability measurements. Gas measurements are quick, as small gas flows can be measured accurately, whereas small liquid flows may have to be collected for a long time to obtain a measurable volume; however, allowances must be made for changes in gas volume caused by expansion and for the fact that gas molecules 'slip' through small-pore channels more easily than do liquid molecules. Gas techniques are also available for the measurement of porosity.

Laboratory tests offer the possibility of making accurate measurements under carefully controlled conditions on samples of precisely known geometry, but the samples are inevitably small. To test more representative volumes we must leave the laboratory and move 'into the field'.

Field measurements

One of the most effective and frequently used methods of measuring aquifer properties is the **field pumping test**. We saw in Chapter 6 that when water is pumped from a well, a cone of depression is formed in the potentiometric surface. The steepness of the cone depends on the hydraulic gradient, which in turn depends on the pumping rate and on the transmissivity and storage coefficient of the aquifer. The storage coefficient relates the volume of the cone to the total quantity of water pumped out; the smaller the storage coefficient, the larger the cone must be for any given quantity of water abstracted. Knowing these relationships it follows, in the absence of complicating factors, that if we pump water from a well and observe the way the cone of depression expands, then we should be able to deduce the transmissivity and storage coefficient of the aquifer. This is the principle of the pumping test, but the phrase 'in the absence of complicating factors' must be kept in mind.

The usual procedure for a test is that water is pumped from one well – called the **production well** or **pumped well** – at a constant rate, which is carefully controlled and measured. The resulting change in the potentiometric surface is monitored by measuring the change in head in one or more **observation wells** near the pumped well. (Provided that the density of the groundwater is constant, the change in water level in a well is an accurate representation of the change in head.) A possible arrangement of wells for a test is shown in Figure 10.5. The longer pumping

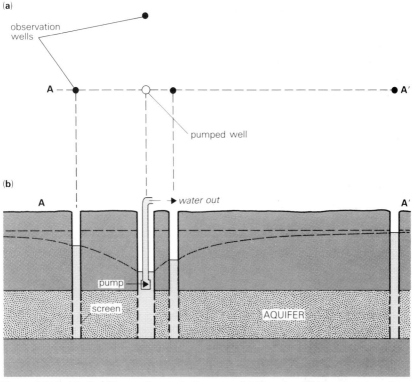

Figure 10.5   Idealised layout for a pumping test   (a) Plan view. (b) Section along line A–A'.

continues, the further the cone of depression will expand. At greater pumping rates the hydraulic gradient must be steeper and the cone must therefore be deeper, causing larger drawdowns; however, simply increasing the rate of pumping will not cause the cone to extend further outwards.

In a perfectly confined aquifer (one that has totally impermeable confining beds), the cone of depression will continue expanding until it meets some form of boundary. The growth of the cone will be rapid at first, but will become slower, because each time the radius of the cone is doubled the volume is quadrupled; as the pumping rate is constant, the *volume* of the cone must increase uniformly and its *radius* must therefore grow ever more slowly. The cone must grow at a finite (albeit slow) rate until it intersects some form of recharge.

If the potentiometric surface was initially horizontal, and if the aquifer is perfectly homogeneous and isotropic, the cone of depression will be symmetrical about the pumped well. For these conditions Charles V. Theis, of the USGS, produced in 1935 an equation relating the drawdown of the potentiometric surface at any distance from the production well to the transmissivity and storage coefficient of the aquifer and the rate and duration of pumping. Because the cone of depression is assumed to be symmetrical, only one observation well is needed, in theory; however, if only one observation well is used, it is impossible to tell whether or not the required symmetry is present. In practice, it is therefore advisable to use at least two observation wells as well as the production well; measurements of drawdown are made at frequent intervals in all wells.

The nature of the Theis equation makes it easy to calculate the drawdown when the properties of the aquifer and other factors are known, but difficult to calculate the aquifer properties when the drawdown is known. The usual way of using the equation is therefore to plot a graph of drawdown against time for each well on double logarithmic (log–log) graph paper, and to match these data plots to a 'type-curve' drawn on similar paper. From the relative positions of the graph scales, the transmissivity and the storage coefficient can be determined. If several observation wells are used, data from them all should yield roughly the same values of transmissivity and storage coefficient; if they do not, the cone of depression is not symmetrical and the Theis equation is not strictly valid. If no observation well is available (this sometimes happens in the early stages of a groundwater investigation, or when testing very deep aquifers) then measurements in the production well make it possible to determine the transmissivity but not the storage coefficient.

Theis derived his equation by using an existing solution for an analogous problem in heat flow. It was a tremendous step forward, because until its derivation the commonly used pumping-test formula was one developed in 1906 by G. Thiem in Germany from earlier work by the Frenchman, J. Dupuit; their equation describes the flow of groundwater to a well under equilibrium conditions. These conditions exist only if the aquifer is receiving uniform recharge around the cone of depression, although in practice they are approximately satisfied after long periods of pumping, when the cone of depression is expanding very slowly. Because it assumes that the cone of depression has stopped expanding and water is no longer being taken from storage, the Dupuit –Thiem equation cannot provide a value for the storage coefficient.

The Theis equation therefore ranks as of fundamental importance in hydrogeology. Various approximations allow it to be used graphically without the need for type curves, and it can also be used with data from the recovery of the potentiometric surface at the end of pumping. Other workers have produced techniques or modifications for use when the aquifer does not conform to the strict requirements of the Theis equation, namely that the aquifer is perfectly confined, has an infinite horizontal extent, is homogeneous with constant transmissivity in space and time, has an initially horizontal potentiometric surface, and releases water from storage instantaneously as the head is reduced; and that the pumped well is of negligible diameter, completely penetrates the aquifer, and is pumped at a constant rate.

Theis himself suggested that some of the assumptions were more important than others. For example, if the diameter of the well is large, it takes a finite time for the water in the well to be removed; the effect is seldom important in practice except in dug shafts or in very deep wells, or where the aquifer is of low permeability; in all these cases, the water in storage in the well may be equal to many minutes' worth of pumping. 'Infinite horizontal extent' means, in practice, that the cone of depression never reaches a boundary; if it does, a recharge boundary will reduce the rate of drawdown and a barrier boundary will increase the rate of drawdown.

If the aquifer is not perfectly confined, water may leak through the confining beds into the aquifer in response to the reduction in head caused by pumping. This leakage, like a recharge boundary, will reduce the rate of drawdown by supplying additional water; Mahdi Hantush produced the first equations to deal with this problem.

When water is pumped from a confined aquifer, the aquifer usually remains fully saturated; only the water pressure is reduced. In an unconfined aquifer (Fig. 10.6a) withdrawal of water means that the water table

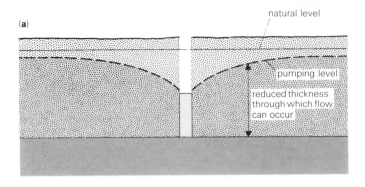

(a)

natural level

pumping level

reduced thickness through which flow can occur

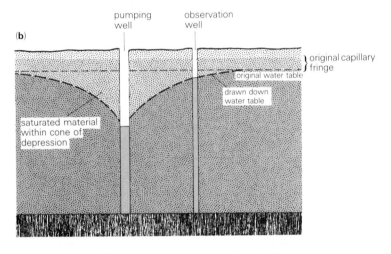

pumping well

observation well

(b)

} original capillary fringe

original water table

drawn down water table

saturated material within cone of depression

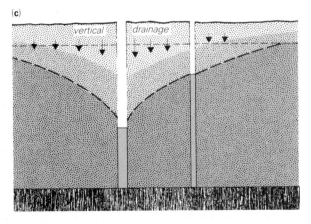

(c)

vertical   drainage

Figure 10.6   The cone of depression in an unconfined aquifer   (See text.)

(d)

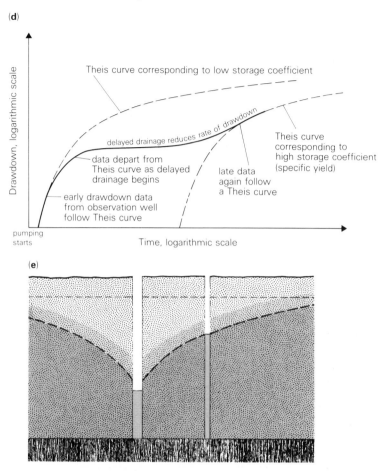

(e)

is lowered, and the saturated thickness of the aquifer is reduced. This alone can cause problems in analysis; however, as Theis pointed out, if the drawdown is small in relation to the saturated thickness of the aquifer, the resulting change in transmissivity may be small. A more important problem is the way in which the water is released from storage. When pumping starts, the head around the pumping well declines rapidly, as it does in a confined aquifer; the water table, which is the level at which the water is at atmospheric pressure, is therefore lowered. As in a confined aquifer, the first water released comes from elastic storage as the aquifer compacts and the water expands slightly in response to this reduction in pore-water pressure. Initially, however, little or no water drains from the pore space. We therefore have the condition shown in Figure 10.6b; the water table has been lowered to form the cone of depression, but within this cone most of the material is still fully saturated – in effect, we have a thick conical capillary fringe. For a short time after pumping starts, the drawdown curve therefore follows that predicted by the Theis equation (Fig. 10.6d).

The thickness of saturated material in the cone of depression is generally too great to remain saturated, however – the capillary columns are too long in relation to their diameters – and so this material begins to drain vertically towards the water table (Fig. 10.6c). This drainage has the effect of recharge entering the cone of depression: it causes the rate of drawdown to be less than that predicted by the Theis equation (Fig. 10.6d).

As pumping continues the cone of depression expands more slowly, and the drainage of the pores within the cone is able to catch up and keep pace with the growth of the cone (Fig. 10.6e). The time–drawdown behaviour of the aquifer (Fig. 10.6d) returns to the curve predicted by the Theis equation, but the curve is displaced from the starting point. The complete response of an unconfined aquifer to pumping can thus be thought of as comprising three stages: initially, the response is like that of a confined aquifer, with water being released from compressible storage; in the second stage, the rate of drawdown decreases as vertical drainage takes place from saturated material within the cone of depression; then, as this drainage catches up with the growth of the cone, the drawdown again follows a Theis curve, but one corresponding to a higher storage coefficient (the specific yield). The time for the second stage (which is known as the **delayed yield** stage) to be completed can range from hours to weeks or more; it will be prolonged if drainage is delayed by material of lower vertical permeability or by capillary effects.

The reduction in transmissivity caused by reducing the saturated thickness of the aquifer can cause further complications, as can the vertical flow components within the saturated part of the cone; the response of unconfined aquifers is therefore complicated. The first person to produce equations describing the three-part nature of the response was N. S. Boulton in England; the understanding of unconfined behaviour was greatly advanced by S. P. Neuman in the 1970s. These advances, like many more in pumping-test analysis, relied on the availability of computers to handle the numerical calculations involved for the many different cases. It is worth noting, however, that an essentially correct qualitative description of a cone of depression in a water-table aquifer was given in 1850 by Robert Stephenson, the famous civil engineer (and son of the even more famous 'father of the railway', George Stephenson). In a report on the water supply of Liverpool, Stephenson not only described the shape of a cone of depression, but correctly surmised the way in which a geological fault would act as a barrier to groundwater movement – all this, six years before Darcy published his work.

Most hydrogeologists would agree that the pumping test is generally the most satisfactory way of measuring aquifer properties. Unlike laboratory techniques it tests a significant volume of the aquifer, in place, whether the flow through it is predominantly through pores, through fissures, or through a combination of the two; it does this, in addition, using formation water. When something seems too good to be true, we are right to be suspicious!

To begin with, compared with laboratory measurements, what the field test gains in volume tested it loses in precision and control. No longer is the tested sample of carefully measured size, held in a permeameter; it is an uncertain volume of aquifer, whose thickness and lithology can be precisely ascertained only where expensive boreholes have been drilled. And an aquifer that looks uniform on the basis of information from two or three small-diameter boreholes can contain some large surprises. One of the beauties of the pumping test is that if it continues for long enough the aquifer properties deduced from it will be useful averages of the minor variations that are present; the problem comes in deciding when it has continued for long enough. A test may, by chance, be ended after a 'minor variation' alone has been studied, and this fact might not be discovered until expensive development boreholes have been drilled in the wrong places or to the wrong depths.

If one knew all the variations in the shape of the cone of depression in both space and time, it might be possible to produce a unique and accurate description of the hydraulic variations that the cone has encountered. Unfortunately, we cannot achieve this complete knowledge: the only points at which we know the position of the potentiometric surface are the observation wells. Pumping tests do not therefore avoid the problem of sampling: we are effectively 'sampling' the cone of depression when we make measurements in the observation wells. In many tests, several sets of conditions could produce the same 'sampled' results.

Finally, we have to ask whether the aquifer is undisturbed by the testing. Probably it is not. As we have discussed earlier (Figs 6.10c and 8.1, for example), there are vertical components of flow – and therefore vertical hydraulic gradients – in most aquifers. These gradients can cause significant differences in head between points with different elevations; the presence of layers with low permeability will tend to cause these differences to be increased. Boreholes can provide highly permeable pathways between these points with different heads, creating what in electrical terms would be a 'short circuit' and disturbing the very parameter they are intended to measure.

The combined approach to hydraulic measurements
A cynic might summarise the position by saying that laboratory tests permit accurate measurements to be made on disturbed, unrepresentative material; whereas field methods provide rather inaccurate measurements on what was representative material only before the test boreholes were drilled through it. Even to the least cynical, it may appear by now that aquifers are so variable, and the methods for studying them so imprecise and beset with problems, that the prediction of groundwater behaviour requires a crystal ball rather than a computer. Most hydrogeologists would probably agree, at least in part, with such a conclusion; nevertheless the predictions must be made and reliable crystal balls are hard to find. What should the hydrogeologist do? Field studies are expensive and unlikely to provide a unique interpretation; the apparent precision of results from laboratory methods depends on which samples were chosen, and cannot take account of fissures or aquifer boundaries. If time and money are available, the investigator will probably choose somewhere between these two extremes. Just as an understanding of the hydrogeology of an area involves studying the geology, hydrology and groundwater hydraulics, so the study of the groundwater hydraulics can itself be properly achieved only by employing a variety of techniques.

How far it is necessary or possible to pursue this type of combined investigation will depend on circumstances such as how much is already known, and whether the investigation is for a major scheme involving many wells over a large area or for a single production well. It will depend too on how quickly the results are needed, and on how much money and how many people are available for the work; these last circumstances are usually beyond the control of the hydrogeologist. Scientists can make recommendations, but the decisions are usually made by politicians, and, unfortunately, are sometimes made for political rather than technical reasons. Thus it may happen that large schemes, on which votes or prestige may depend, are rushed through with less time for investigation than smaller ones; those in authority often feel that a job which should take four years should be possible in one year if four times as much money is made available. This sort of attitude overlooks the fact that trained and experienced workers cannot be obtained at the drop of a hat, that exploration programmes must progress in a phased way if money is not to be wasted, and that several years of data collection may be necessary for proper plans to be made.

Leaving aside political and economic constraints, an investigation of the groundwater hydraulics of an area of hundreds of square kilometres might commence with a series of exploratory boreholes, some of which would be core drilled if possible and all of which would be logged. The holes could then be used for measurement of water-level fluctuations and natural hydraulic gradients. Preliminary pumping tests would be carried out to obtain some idea of the transmissivity of the aquifer or aquifers. Also at this stage samples of water would be taken for chemical and biological analysis (Ch. 11). If all the results were encouraging, a further pumping test or tests would be undertaken with observation wells.

At least some of the cores would be subjected to laboratory measurement of permeability, and the results compared with those obtained from the pumping tests. Frequently there are discrepancies between the two, the presence of fissures commonly resulting in field values higher than the corresponding laboratory values.

Geophysical logging of the borehole fluid column is an excellent way of detecting or verifying the presence of fissures. The temperature and chemical quality of groundwater usually vary slightly from place to place and with depth, even in a single aquifer. The column of water in a borehole usually represents an average of the groundwater in the surrounding aquifer, and at any particular level in the aquifer the properties will usually differ slightly from this average. If water enters the borehole in relatively large quantities from a particular level – because of the presence of a transmissive fissure or of a layer of highly permeable rock – there will thus be a slight change in the temperature and chemical quality of the borehole water at that level. These changes can be detected using sensitive temperature and electrical-conductivity logging equipment, even though the temperature change, for example, may be only 0.01 °C. Figure 10.7a shows an actual example of how an electrical conductivity log indicates a fissure.

Other fluid logging instruments include a variety of **flowmeters**, which measure the speed of flow of water up or down a borehole. The movement may result from natural head differences or be the result of pumping; major changes in speed are associated with levels where water enters or leaves the borehole – again these are fissures or highly permeable layers (Fig. 10.7b).

Finally, what is in some ways the simplest technique (though it requires sophisticated equipment) is to look at the borehole wall using an underwater closed-circuit television camera, with a television monitor on the surface. Fissures can be observed directly and their openings and orientation estimated. The results can be recorded on videotape or the screen can be photographed (Fig. 7.7).

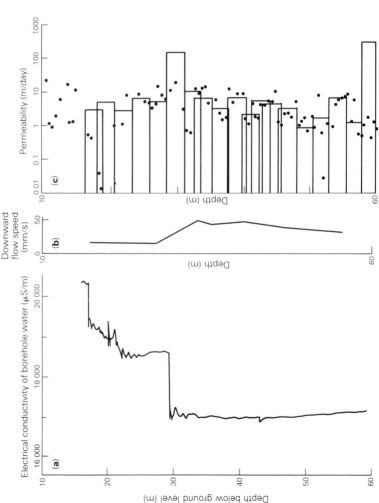

Figure 10.7 Geophysical and permeability measurements in a borehole in Permian sandstone (a) An electrical conductivity log. (b) A flow velocity log. Both (a) and (b) indicate a major inflow of water at about 29.5 m below ground level. (c) A comparison of intergranular permeability measured in the laboratory, indicated by dots, and total permeability measured using packer tests, indicated by rectangles. (The right-hand end of each rectangle indicates the average permeability for the depth interval represented by the rectangle.) There is generally good agreement between the two sets of measurements except at about 30 m and 60 m where television inspection revealed major fissures – the higher fissure caused the inflow detected by the logs in (a) and (b). (Reproduced by permission of the Director, British Geological Survey.)

Combining all these techniques, the hydrogeologist can either obtain a reasonably consistent picture of the aquifer or, at worst, see where further study is needed.

To obtain a quantitative estimate of the contribution which particular layers or fissures make to transmissivity at a site, sections of boreholes can be tested individually by isolating them between inflatable packers (rubber sleeves which are expanded to seal against the borehole wall) and

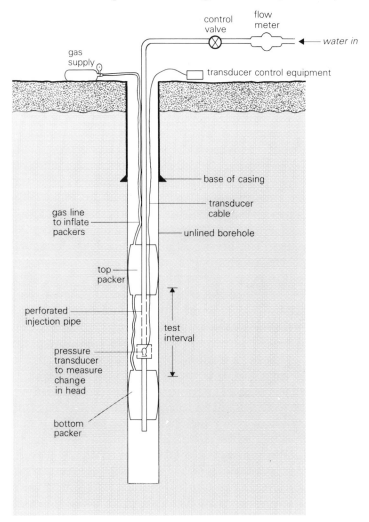

Figure 10.8 Arrangement for testing a borehole interval isolated between packers Water is pumped into the interval at a measured rate, and the resulting head change is measured with the transducer.

pumping water into or out of the isolated section (Fig. 10.8). An example
of the results obtained with this method, and how they compare with
laboratory measurements and flow logs, is shown in Figure 10.7. This
shows that consistent results can be obtained if sufficient trouble is taken
to collect the data.

Models

Having gone to some trouble to collect geological, hydrological and
hydraulic information about an area, what do we do with it? Usually the
information is needed so that the water resources of the area can be
evaluated, calculations made of the rate at which water can be
abstracted, the effects of abstraction predicted, and so on. Before these
calculations can be made, a **model** of the aquifer is needed.

At its most abstract, a 'model' is a unification of the concepts about
the aquifer which the hydrogeologist uses in making his predictions. For
example, he may have concluded from his investigation that the aquifer
is confined, has high intergranular permeability, and is traversed by
numerous fissures which cause local variations in aquifer properties but
which, on a regional scale, increase the transmissivity in a uniform way.
Also, the aquifer extends well beyond the study area. As a first approx-
imation, he may therefore be justified in treating the formation as an infinite, homogeneous, isotropic confined aquifer to which he can apply
the Theis equation to predict drawdown around pumping wells; this
would be his conceptual model of the aquifer. If he found that the fissur-
ing was restricted mainly to the upper part of the formation, causing a
markedly higher permeability there, he would probably change his idea
to a two-layer model. He can use these models, or ideas, as the basis for
manual or graphical predictions of, say, drawdown resulting from the
simultaneous pumping of several wells.

As the number of wells increases or the aquifer conditions or geometry
become more complex, it becomes more difficult and time-consuming to
do the calculations, even if the equations can be solved. A model is then
needed that is more than just an idea.

One possibility is a *physical* model of the aquifer, which usually takes
the form of a **sand-tank model**. In this model the aquifer is represented
by a layer or layers of sand in a watertight tank. Impermeable beds or
boundaries can be sheets of plastic, and the potentiometric surface is
observed in thin tubes which act as observation wells or piezometers.
Other tubes serve as production wells. A major problem is that capillary
effects tend to be large in relation to the size of the tank, making it

difficult to model regional unconfined aquifers; however, sand models have been widely used to study flow near wells.

Another type of model, offering more scope for regional simulation, is the **analogue**. This uses the analogy between the laws of groundwater movement and the laws governing the movement of some other physical quantity. One of the most versatile analogues, the **resistance network**, uses the similarity between the flow of groundwater and the flow of electric current. (See the box on p. 48.)

In resistance analogues a model of the aquifer is constructed using electrical circuit components. Usually the models are built on a grid basis with electrical components meeting at junctions called **nodes** which correspond to the intersection of grid lines on a map of the aquifer (Fig.

Figure 10.9   Electrical resistance analogue model of an aquifer   (Reproduced by permission of the Director, British Geological Survey.)

10.9). The resistor connecting two nodes is chosen to have a resistance that is inversely proportional to the transmissivity of the aquifer between the corresponding nodes on the map. The electrical potential (voltage) at any node represents the head in the aquifer at the corresponding point, and electric current represents flow of water. The model conforms to the shape of the portion of aquifer being considered, and appropriate conditions of current and potential must exist at the boundaries. If flow is not steady, storage has to be considered; in the model this is achieved by connecting capacitors, which store electricity, to each node. The capacitance of the capacitor is proportional to the storage coefficient of the aquifer at the corresponding point.

The grid is drawn in such a way that the size of the square on the ground is small in relation to the area of aquifer being studied. It is not practical to measure the aquifer properties in every grid square; usually they are measured in a few places and all other values are estimated or interpolated. The model is **calibrated** using historical data; observed values of head are fed into the model as starting conditions, and the model's prediction of the head changes after a suitable interval is compared with what really happened. The analogous 'head' (the potential) is measured with some form of voltmeter. The components used in the model are chosen in such a way that processes which take several years in the aquifer are simulated in the model in seconds.

If the predictions from the model do not agree with the historical records, the resistance or capacitance values probably do not accurately represent the aquifer properties. They must be altered, by changing the components, until agreement is achieved.

Once the behaviour of the model agrees with the behaviour of the aquifer, it can be used to predict the aquifer's response to changing conditions. Pumping can be simulated, for example, by using an electronic signal generator to produce the appropriate voltage fluctuations at the node corresponding to the production well.

Electrical analogues were popular in the 1960s. Their main disadvantage is that they are laborious and time-consuming to construct and modify during calibration, and that they can simulate the behaviour of only one aquifer. They are usually constructed as two-dimensional models, with no simulation of layering or vertical flow. In the 1970s their place was increasingly taken by **digital models**, which make use of the increasing power and availability of digital computers; they are in essence computer programs.

From the mathematical point of view, the simplest type of digital model is the **finite difference** model. This also involves dividing up the aquifer into smaller units; they need not be the same size and shape, but squares are often used for simplicity. The aquifer properties for each square and the initial heads in the squares are fed into the computer, which then calculates the flow that will occur over a short time interval and what the heads will be at the end of that time. It repeats the process as many times as required, taking into account information on recharge and abstraction.

As with analogues, a digital model must be calibrated using historical data before it can be used to make predictions. The advantage of the digital model is that changes found to be necessary during calibration require only a new set of values to be fed into the computer, not physical alterations to the analogue. Similarly, the same model can be used to study different aquifers by changing the aquifer properties and boundary conditions.

In the manner of Parkinson's law, models expand to fill the computer storage available. Desk-top computers available today will handle models which, ten years ago, would have needed a mainframe computer, but modellers keep developing models of greater sophistication which use to the full the storage available on the newest computers. These models will simulate the behaviour of aquifers of complicated shape and variable lithology and can be used to predict changes in water chemistry. The use of these models can often reveal flaws in existing knowledge of the aquifers. However, discretion has to be exercised in their use; the phrase 'rubbish in, rubbish out' indicates the predictions that can be expected from a model, however sophisticated, which is fed with unreliable data or which is based on incorrect suppositions.

Water divining

No discussion of groundwater exploration would be complete without a reference to **water divining** (also called **dowsing** or **water-witching**). This is the method by which some people claim to be able to locate ground-water by walking over the surface of the ground until they observe a response with a forked stick, bent rods, a pendulum or some other apparatus which is usually held in front of them with both hands.

It is difficult to assess the technique objectively; there is at present no scientific explanation as to why it should work. Moreover, whenever the technique has been tested impartially, the success of the proponents has been no more than would be expected from chance. A dowser can walk across the English Chalk and predict that water will be found below a certain spot; a hydrogeologist knows that a well drilled almost anywhere on the Chalk will encounter water. However, the subject cannot simply be dismissed. Many people (including some scientists and engineers) can locate buried pipes with the aid of rods or twigs. One theory is that muscles in the body react to some electromagnetic effect caused by the presence of the metal or of the flowing water in the pipe; the rods amplify these minor 'twitches' so that the searcher is aware of them. Another theory is that after long experience in an area, some diviners know intuitively the likely settings in which groundwater will be found near the surface, and subconsciously cause the reaction.

There may or may not be something in these explanations. Even if the electromagnetic theory works for pipes, there is no reason why it should detect the slow, diffuse movement of groundwater in an aquifer.

Perhaps the most telling argument against the use of divining however is that simply locating the presence of groundwater is but a small part of the task of hydrogeology. The hydrogeologist must consider the long-term effects of abstraction, whether it will affect other wells or springs, whether there will be changes in quality with time, and so on. It is difficult to see how the diviner's twig can deal with all these problems.

Selected references

Freeze, R. A. and J. A. Cherry 1979. *Groundwater*. Englewood Cliffs, NJ: Prentice-Hall. (See especially Ch. 8.)

Griffiths, D. H. and R. F. King 1981. *Applied geophysics for engineers and geologists*. Oxford: Pergamon.

Kruseman, G. P and N. A. De Ridder 1983. *Analysis and evaluation of pumping test data*, 3rd edn. Bull 11. Wageningen, Netherlands: International Institute for Land Reclamation and Development.

Price, M., B. Morris and A. Robertson 1982. A study of intergranular and fissure permeability in Chalk and Permian aquifers, using double-packer injection testing. *J. Hydrol.* **54**, 401–23.

Tate, T. K., A. S. Robertson and D. A. Gray 1970. The hydrogeological investigation of fissure flow by borehole logging techniques. *Q. J. Engng Geol.* **2**, 196–215.

Theis, C. V. 1935. The relation between the lowering of the piezometric surface and the rate and duration of discharge of a well using ground-water storage. *Trans. Am. Geophys. Union* **16** (Part II), 519–24.

Todd, D. K. 1980. *Groundwater hydrology*, 2nd edn. New York: Wiley.

11 Water quality

Water for drinking

In 1854 a cholera epidemic struck Soho, in Central London, causing an estimated five hundred deaths in ten days. Unfortunately, such events were not uncommon around that time; between 1848 and 1850, more than 50 000 people died in a cholera epidemic in Britain.

Central London in the middle of the 19th century was an insalubrious place, with many people supplied with water from the River Thames which also received their untreated sewage. Overcrowding was appalling, and it was commonly believed that diseases like cholera were spread by 'noxious vapours'. A contemporary book went so far as to publish a 'cholera map', which correlated the incidence of cholera with elevation of the land; the incidence of cholera was greatest on the low-lying areas near the river, which were subject to damp and mists. Therefore, went the argument, cholera was clearly caused by the 'mists' or 'vapours'.

One person who had doubts about the 'vapours' theory was an anaesthetist called Dr John Snow. He had pioneered the use of chloroform in place of ether in anaesthesia, and had achieved recognition in his profession by administering chloroform to Queen Victoria at the birth of two of her children. Snow noticed that the low-lying areas in which cholera was most prevalent were all served by two water companies which drew water from the Thames; the higher, healthier areas were supplied from other sources. He reasoned that it could be water, not 'mists', which carried cholera, and the epidemic of 1854 was to give him the proof he needed.

Faced with the spread of the disease Snow did something which today might seem obvious, but at that time was almost revolutionary – he plotted each case of cholera on a map. The points clustered around a hand pump in Broad Street in Soho, which drew water from a well in the terrace gravels of the Thames, and Snow was able to verify that almost all of the people affected by the disease had drunk water drawn from the well. Two cases stood out – these were at a house in Hampstead, 6 km away. Snow found that the house was occupied by a widow who had moved there from Soho. She found that the Hampstead water was not to her taste, and so sent a servant each day to collect the familiar water from Broad Street – a small indulgence which cost her own life and that of her niece.

For Snow, his map and discoveries about the well were the final pieces of evidence. He went to a meeting of the Board of Guardians of St James' Parish who, on 7 September 1854, were discussing the epidemic and seeking some way to combat it. Snow's advice has passed into legend as 'Take the handle off the Broad Street pump'. This advice was followed, the handle being removed on the following day. By Snow's own account, published the following year, the epidemic had been declining anyway – by then, the majority of the local people had either died or fled – so it is difficult to say categorically that his discovery halted the epidemic. What it did do was to make the authorities aware of the way in which cholera was transmitted, years before cholera bacteria were identified, and in that respect Snow's work probably saved thousands of lives.

(The Broad Street pump has long vanished, and Broad Street is now Broadwick Street. At the junction of Broadwick Street and Lexington Street, close to the spot where the pump stood, a public house has been renamed the 'John Snow'. It contains a copy of Snow's map, and other information connected with the epidemic.)

Cholera is now known to be caused by bacteria which enter the intestinal tract of infected individuals and then multiply rapidly. The excreta of these individuals contain cholera bacteria which, in the absence of adequate sanitation, can be carried into watercourses. If this water is drunk without being boiled or treated other people become infected. The Soho epidemic was caused by untreated sewage entering the shallow groundwater of the terrace gravels.

Other intestinal diseases which are caused by water-borne bacteria include typhoid, bacillary dysentery and some forms of diarrhoea; all of these diseases are spread in much the same way as cholera. Amoebae can cause other forms of dysentery. Infectious hepatitis (jaundice) is a virus disease that can be spread by water (although it can be transmitted in other ways) and other virus diseases are possibly spread in the same way (poliomyelitis is almost certainly one of them). To prevent the spread of these diseases, water supplies must be tested before they are used by public undertakings.

A single virus can be infective to humans (whereas at least ten bacteria would be needed), and viruses are too small to be seen with an optical microscope. A further complication is that viruses can reproduce only in living tissue and they are specific as to the type of cells they kill, so that tests have to be done on appropriate living cells – human cancer cells and monkey kidney cells are among the most commonly used. As the viruses reproduce, they cause a patch or **plaque** of dead cells 1 mm to 2 mm across, and the concentration of viruses in water is expressed in terms of the number of **plaque-forming units** or **PFUs** found in a litre of water.

In practice, because it takes a long time to locate and identify disease-producing organisms in contaminated water, a simpler technique of testing is adopted. All humans have bacteria of a species called *Escherichia coli*, one of a group known as **coliform bacteria**, living in their intestines. The presence of coliform bacteria in water implies that the water *may* have been contaminated with human excreta, which *may* have contained disease-producing (**pathogenic**) organisms. In practice, therefore, the bacterial examination of water is usually limited to a search for coliform bacteria.

Although some species of bacteria appear to flourish in some aquifers, groundwater is usually free from pathogenic micro-organisms. The natural filtering action of the aquifer on the percolating water, combined with the long residence time of water in the aquifer, are usually sufficient to render the water biologically safe. Only if the water moves rapidly through relatively large openings such as fissures or the pores of coarse gravel, or if there is a source of pollution close to the well, or if the well construction permits the entry of contaminated surface water, is abstracted groundwater likely to contain pathogenic organisms. One or more of these factors was probably responsible for the Broad Street epidemic.

Since 1937, when a typhoid epidemic in Croydon was attributed to contaminated water from a Chalk well, all public water supplies in Britain have been required by law to be disinfected. This is usually done with chlorine at a dose of one milligramme of chlorine per litre of water; groundwater usually requires little else in the way of treatment. The filtration treatment given to surface water supplies usually happens naturally in aquifers.

Because it is less likely to contain pathogenic organisms, groundwater is much more suitable than surface water for supplies of drinking water in isolated areas, where the treatment undertaken in large-scale supply systems is not possible. Provided that proper precautions are taken in drilling and completing wells, groundwater can usually be drunk with no treatment at all. This is a particular advantage in providing supplies in developing countries, where even the most elementary form of water treatment – that of boiling the water for a few minutes to kill the organisms present – is beyond the scope of many of the inhabitants, who cannot afford the necessary firewood or other fuel.

Physical, chemical and biological aspects of quality

Water from any source must suit the purpose for which it is intended. Water intended for drinking must, as outlined above, be free from pathogenic organisms; in other words, it must have suitable *biological* quality. Its *chemical* and *physical* quality must also be suitable. In chemical terms, it must not contain any dissolved or suspended material that would be injurious to health or that would give it an unpleasant taste. Physically it must be at a suitable temperature and not have objectionable colour or cloudiness.

Water for irrigation need not meet such strict biological requirements (although it is bad practice to distribute polluted water anywhere where people may come into contact with it), but chemical quality may be important. Industrial requirements vary considerably; some processes need only poor-quality water, while in others the presence of suspended or organic material can block filters, and dissolved material may cause scale formation or may accelerate corrosion. Some processes have their own problems; minute amounts of iron or manganese can be disastrous to a laundry, for example, because they cause staining.

Groundwater chemistry

The study of the chemistry of groundwater is a vast topic which ranges from the routine (but essential) work of the water-supply chemist, who must ensure that the water supplied by his undertaking meets the requirements of consumers and statute, to the esoteric studies of the geochemist concerned with diagenetic changes and the slow interaction between groundwater and its host rock.

Most groundwater abstracted for domestic, industrial or agricultural use is **meteoric** water, i.e. groundwater derived from rainfall and infiltration within the normal hydrological cycle. The word meteoric comes from the same root as 'meteorology' and implies recent contact with the atmosphere. As we shall see later, the chemistry of meteoric groundwater changes during its passage through rocks, the changes depending on such factors as the minerals with which it comes into contact, the temperature and pressure conditions, and the time available for water and minerals to react. The modification of meteoric groundwater in its passage into and through the ground is one of the evolutionary sequences of groundwater chemistry, and it occurs in many aquifers.

The generally saline waters encountered at great depths in sedimentary rocks were at one time all thought to have originated as sea water trapped in marine sediments at the time of their deposition, and were referred to as **connate** waters. It is now accepted that meteoric waters may eventually become saline. It is also accepted that most of the original sea water has been modified and has moved from its original place of entrapment; this evolution of marine waters is the second sequence of change of groundwater chemistry. There is now some debate as to whether the term 'connate' should be used to refer to entrapped marine water in its original sediments, to marine water that may have migrated, or to any saline groundwater; or whether indeed it should be used at all. Most workers would agree that the term 'connate' implies that the water has been removed from atmospheric circulation for a significant (in geological terms) length of time; I shall use it to denote groundwater derived mainly or entirely from entrapped sea water.

There is a third possible origin for groundwater mineralisation. **Juvenile** water is the relatively tiny amount of water believed to be derived from igneous processes within the Earth; it has never previously taken part in the hydrological cycle and can contribute unusual constituents to the meteoric groundwater which it joins. A problem for geochemists is that juvenile water is indistinguishable, with present techniques, from meteoric water that has penetrated to great depths and become intimately associated with igneous processes.

Chemical development of meteoric water

Meteoric water, by definition, was once precipitation. Although rain is nature's form of distilled water, it typically contains between 10 and 20 mg/l (milligrammes of solute per litre of solution) of dissolved material. Near coastlines the concentration of sodium chloride is increased, and downwind of industrial areas sulphur and nitrogen compounds are more in evidence. The presence of these compounds turns the rain into dilute acid, and in some parts of Europe and North America **acid rain** is viewed as a major environmental problem.

When the precipitation infiltrates the soil, the most important natural change is the dissolution of carbon dioxide from the soil atmosphere. The weak carbonic acid so formed is then able to dissolve calcium carbonate, if any is present in the underlying rocks. The soil organisms also consume much of the oxygen that was dissolved in the precipitation in its passage through the atmosphere.

In temperate and humid climates, where recharge is a regular process, water is usually moving relatively rapidly through the outcrop area of an aquifer, so its contact time with the rocks may be limited. Any highly soluble materials, such as sodium chloride, will have been flushed from the system long ago, and there will often be insufficient time for poorly soluble minerals to be taken into solution in significant amounts. As a result, groundwater in the outcrop areas of aquifers in countries like Britain tends to be low in dissolved solids unless the aquifer is a limestone or a sandstone with calcium-carbonate cement, in which case the groundwater will contain calcium and bicarbonate as the dominant ions.

Sulphate and nitrate will also be present in solution in small quantities. Nitrates are usually derived from the soil, where they may originate from the fixation of atmospheric nitrogen by leguminous plants or from the oxidation of organic materials by bacteria; additionally, large quantities of nitrates are nowadays being leached from agricultural fertilisers. Sulphate ions are commonly produced in the outcrop area by the oxidation of metallic sulphides that are present (at least in small quantities) in many rocks; the oxidising agent is atmospheric oxygen which was dissolved in precipitation. The common sulphate minerals – gypsum and anhydrite – are readily soluble, and like sodium chloride will usually have been flushed from the outcrop area already.

If the aquifer dips below a confining bed, the conditions prevalent at the outcrop may continue for some way below the impermeable cover. Dissolution is still the dominant process, and in aquifers containing calcium carbonate the amounts of calcium and bicarbonate in solution typically rise to their highest levels in this part of the aquifer.

With increasing distance from the outcrop, the dominant process changes from dissolution to ion exchange. Most aquifers contain some clay minerals; the small size of clay particles means that, although the clay may be present in only small total quantities, it presents a relatively large surface area to the percolating groundwater. Ions adsorbed on the clay surfaces tend to exchange with ions in solution; this principally involves positive ions (cations), the trend being for the cations in the water to come into equilibrium with those on the clay particles in the aquifer. A major effect is that calcium and magnesium ions in solution are replaced by sodium ions, which are frequently concentrated on clay surfaces when rocks are deposited. The removal of calcium ions from the water can lead to further solution of calcium carbonate, but the end product tends to be a sodium-bicarbonate water. Since it is calcium and magnesium ions that cause hardness, ion exchange leads to a softer water.

At greater distances from the aquifer outcrop there is typically less natural movement of water because there is generally no outlet for the water as the strata dip downwards, except by leakage through the

confining beds. The slower movement means that there is more oppor-
tunity for less-soluble minerals to be dissolved. The reduced throughflow
also means that soluble minerals such as gypsum may not have been
flushed from the aquifer; with even greater distance from the outcrop and
greater depth of burial, this will even be true of sodium chloride. Thus
there is a change from bicarbonate waters to sulphate waters and finally
to chloride waters, with cations correspondingly changing from calcium
and magnesium to sodium.

To summarise (Fig. 11.1), a complete typical sequence of the evolution
of meteoric groundwater starts with water in which the main anions
(negative ions) are bicarbonates. As the water moves deeper, sulphate
ions increase in importance and become dominant; finally, if the system
is deep enough, chloride becomes dominant. The amount of material in
solution also increases with time and depth. The sequence in which the
various cations become important is complicated by the effects of ion
exchange. A broad generalisation, however, is that as the groundwater
becomes older and deeper the dominant cations change from calcium
and magnesium to sodium.

It must be emphasised that the so-called 'typical' sequence is subject

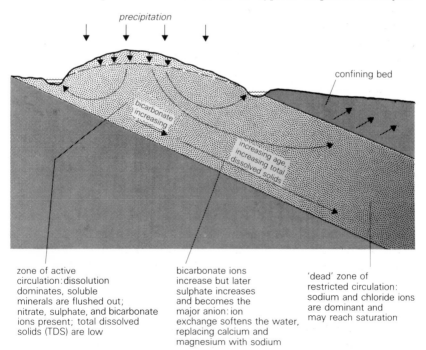

zone of active
circulation:dissolution
dominates, soluble
minerals are flushed out;
nitrate, sulphate, and bicarbonate
ions present; total dissolved
solids (TDS) are low

bicarbonate ions
increase but later
sulphate increases
and becomes the
major anion: ion
exchange softens the water,
replacing calcium and
magnesium with sodium

'dead' zone of
restricted circulation:
sodium and chloride ions
are dominant and
may reach saturation

Figure 11.1   The general chemical evolution of meteoric groundwater

to many variations. For one thing, the water may not stay in the aquifer long enough or move deep enough to reach the 'typical' end product of a sodium-chloride brine. Conversely, if water exceptionally infiltrates an arid region where evaporite deposits are still present, it may reach the sodium-chloride stage almost immediately. The general idea of the sequence is nonetheless a useful one. It is sometimes called the **Chebotarev sequence**, after the scientist who first proposed it.

Another significant change that usually occurs with increasing distance from outcrop (and therefore usually with increasing depth) is the change from oxidising to reducing conditions. As noted previously, groundwater at the outcrop will generally contain oxygen (derived from the atmosphere) and nitrate and sulphate from the soil and outcrop. As the water containing dissolved oxygen, nitrate and sulphate moves away from the outcrop, oxygen is used up in oxidising organic matter and other material such as ferrous iron. Dissolved oxygen is used first, and then nitrate and finally sulphate are reduced as their oxygen is used up. Like the evolution of the major ion chemistry, the time and distance taken for these changes will vary from one aquifer and place to another.

Desert-lain sandstones (such as many of the British Permo-Triassic sandstones) contain very little organic material, and as they were deposited under generally oxidising conditions they have little ability to remove oxygen from groundwater. In these aquifers it is therefore common to find oxidising groundwater at considerable distances from the outcrop area.

In contrast, other aquifers may contain large amounts of organic material, which rapidly depletes the oxidising capacity of the groundwater. In the Lincolnshire Limestone (of Jurassic age), for example, there is a fairly abrupt change from oxidising to reducing conditions about 12 km from the beginning of confined conditions. In this aquifer and in many others, the presence of reducing conditions results, among other indicators, in the presence of hydrogen-sulphide gas.

Groundwater quality in arid areas

The above discussion relates to temperate or humid conditions, where there is an annual surplus of precipitation over evapotranspiration. Aquifer outcrops will usually have been flushed to a depth of several hundred metres or the full thickness of the aquifer by water moving from recharge areas to discharge areas, and the water will generally have a low dissolved-solids content.

In arid and semi-arid areas there may be considerable differences.

Because evaporation or evapotranspiration generally exceeds rainfall for most or all of the year, the aquifer outcrop may be a zone of concentration rather than of dissolution. Unless soluble minerals have been leached from the aquifer in the past, the groundwater may be saline. If the water table is sufficiently close to the ground surface for groundwater to be lost either by capillary action and direct evaporation or by transpiration, there is likely to be a concentration of salts within the soil and subsoil, as water is evaporated or transpired and the solutes are left behind. In this way a crust or 'hardpan' can form at or near the surface.

In such conditions irrigation, far from helping, may aggravate the problem. If groundwater is used for irrigation in arid areas, then the high rates of evaporation mean that soluble material in the groundwater will be deposited in the soil. Any occasional natural recharge or excess of irrigation water is liable to dissolve these minerals and carry them down to the water table, thereby increasing the salinity of the groundwater.

In the Indus Valley of Pakistan, irrigation over the past century has caused a rise in the water table, occasionally to the surface. Evaporative losses have therefore increased and large areas of once-fertile land have been damaged by the increased salinity of the soil. One suggested solution is to operate effective surface drainage alongside the irrigation systems, thus ensuring that salts leached from the soil are removed from the area and not recycled into the irrigation water.

The limited recharge in arid and semi-arid regions often means that groundwater has either migrated from outside the area or infiltrated a long time ago (Ch. 8). Either alternative implies long residence times and possibly incomplete flushing of soluble minerals from the aquifer. Sodium-chloride waters are therefore relatively common. A typical

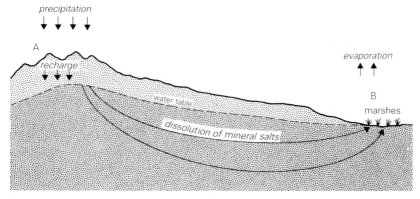

Figure 11.2 Groundwater flow in an arid region In this example, recharge takes place only in the mountains, and discharge occurs in low-lying areas giving rise to salt marshes.

situation might be that of Figure 11.2. Here, recharge on the mountains at A enters and percolates through the aquifer, dissolving soluble materials in its passage, until it emerges (perhaps thousands of years later) in seepages at B. Evaporation is too great to allow the formation of a perennial stream at B, so the discharge area is likely to be stagnant marsh or ponds. Sodium chloride is likely to be the dominant constituent, so salt marshes (like the inland **sabkhas** of North Africa) are a common feature. For the same reason, many of the Saharan oases yield brackish or saline water. Evaporation may lead to the formation of saltpans or beds of evaporite minerals which in the course of time may become part of an evaporite sequence. It is the dissolution of minerals of this type, formed in the geological past, which is in part responsible for the high salinities of some present-day deep groundwaters.

Connate and saline waters

Most meteoric groundwaters abstracted for water supply have total dissolved solids (TDS) concentrations of a few hundred milligrammes/ litre or less. Anything higher would have an unacceptable taste. At depth in some formations the TDS concentration may rise to the level of saturation, which for sodium chloride at typical temperatures is about 250 000 mg/l. The higher concentrations (above a few thousand mg/l) are nearly always dominated by sodium chloride. Since such waters are not abstracted for supply purposes the formations that contain them are not aquifers in the normal sense. Knowledge of the waters is usually obtained from mine drainage, or from oil and gas exploration and production programmes.

Some of these brines may be formed by meteoric water dissolving the salt deposits laid down in earlier geological periods, but many are thought to have originated from sea water trapped in sediments when they were deposited (connate waters). However, since the TDS composition of sea water is at present only about 34 000 mg/l, and it has probably not been significantly higher in the past, some mechanism is needed to explain how the connate water has become concentrated. It seems that the concentration probably comes about by a natural form of reverse osmosis.

If two solutions of similar composition but different strengths are separated by a **semi-permeable membrane** (a membrane that permits the passage of a solvent but not of the dissolved substances) there is a natural tendency for solvent to pass through the membrane from the weaker to the stronger solution. Thus, in the case of two sodium-chloride solutions, water would flow through the membrane to dilute the stronger solution; the weaker solution would become more concentrated and the process would end when the solutions were of the same strength. This process is called **osmosis**. The fact that water flows naturally through the membrane implies that an energy difference must exist as a result of the different chemical concentrations.

Conversely, if the head of the stronger solution is artificially increased (by putting a high pressure on it, for example), solvent can be forced to flow through the membrane from the more concentrated to the less concentrated solution. This process is termed **reverse osmosis** and is used to produce fresh water from sea water in desalination plants. A semipermeable membrane can thus be thought of as a 'sieve' that permits the passage of some ions but not others.

As marine sediments are consolidated by burial beneath other deposits, the sea water they contain would normally be 'squeezed' from them and would generally migrate in an upward direction. However, clays and shales may act as semi-permeable membranes, allowing the water to pass but retaining some of the ions present in the solution. The sieve mechanism actually seems to be electrical in nature: negative charges on the surface of the clay particles repel (and hence prevent the passage of) the negatively charged ions (e.g. the chloride ions). The positively charged ions must also be held back, otherwise the solution would be ionically unbalanced; hence most of the water passes through the clays and the dissolved solids are retained in the pores of their parent sediment. The remaining pore water thus becomes progressively more saline.

Presumably, solutions derived from meteoric waters may become concentrated in the same way, given appropriate conditions.

Isotopes and tracers

It should be apparent by now that the length of time that water spends in contact with the rock, from infiltration to discharge, can have a considerable effect on its quality. Knowledge of this residence time can also be important in assessing present-day recharge and pollution risks; did the water that we pump from a well today enter the aquifer last month, last year or 10 000 years ago? If the last, can we be sure that recharge is occurring now to replenish that water, or are we 'mining' a finite resource? In many places we have endangered the quality of groundwater by dumping poisonous waste materials where leachate from the wastes could reach an aquifer. How long will it take for the toxins to reach an important aquifer, or to pass through a natural flow system to a river?

Groundwater can move through aquifers with speeds ranging from many metres a day to less than a metre a year; this range is possible even within the same aquifer. In confining beds the speeds are, of course, usually slower. How can the residence time of the water be quantified? It would obviously be useful to be able to label a droplet of water with date, time and place, introduce it into an aquifer and await its discharge, much as an ornithologist charts the movements of a ringed bird.

Labels cannot be fixed to water droplets, but for many years cave explorers have used dyes and spores to 'label' the waters of streams that flow through caverns for parts of their courses, and so determine their unseen routes. In principle, the idea can be applied to the study of groundwater; the technique is known as **tracing**, and the substance added to the water to distinguish it is a **tracer**. The substance used for the tracer must move as part of the water, at the same rate, and not become filtered or separated from it; it must not be present naturally in the water (or at least not in the concentrations at which it is being used as a tracer); it must be readily detectable, easy to handle, non-toxic and, ideally, cheap. The fluorescent dyes popular with cave explorers tend to be adsorbed on clays, and the lycopodium spores that the speleologists also use are too dense and too large to be carried by laminar flow through the pores of an aquifer. The ideal tracer, in short, does not exist, but many substances have been used to trace the movement of groundwater and to indicate its natural flow speeds over distances up to a few tens of metres. Once distances increase beyond these, new problems arise: the tracer becomes so dispersed among the bulk of the groundwater that it is likely to be undetectable, and the time taken (except in rapid flow through large fissures) becomes too great — in some cases many times greater than a human lifetime, so that water pumped from a well today would have to have been 'labelled' perhaps thousands of years ago for us to assess its residence time in the aquifer.

Fortunately, nature comes to our aid by labelling water for us; the information that we need is affixed to every drop of water that enters an aquifer — if we can only read the label! The labels that nature provides are 'isotopes'. All the atoms of a given element have the same atomic *number* — their nuclei have the same number of *protons*: however, some have different atomic *mass*, because they have more or fewer *neutrons*. These different forms of the same element are called **isotopes**.

Most elements are mixtures of two or more isotopes. Some isotopes are stable; others (**radio-isotopes**) change by radioactive decay into isotopes of other elements.

The rate of this decay varies from one isotope to another, but for each isotope it is more or less constant and is known fairly accurately. This means that if a radio-isotope of a particular element is present in recharge water in a known ratio to the stable isotope of that element, then during its passage through the aquifer the proportion of the radio-isotope will decrease as that isotope decays. If no more of it is added to the water, it is possible to determine how long it has taken the recharge water to reach any point in the aquifer by measuring the new ratio of radio-isotope to stable isotope at that point.

For example, a radio-isotope of carbon, ^{14}C (carbon-14) is formed continuously by the effects of cosmic rays on nitrogen in the upper atmosphere. Until relatively recent fossil-fuel burning, and the testing of thermonuclear weapons which began in 1952, the proportion of ^{14}C to stable carbon present in the atmosphere as carbon dioxide was believed to have been more or less constant for tens of thousands of years, the decay of ^{14}C being balanced by its production. Carbon in rain and in soil water (as dissolved carbon dioxide) is present in the same isotopic proportions. Once this water enters an aquifer it is no longer in contact with the atmosphere. The ^{14}C decays, so that the proportion of ^{14}C to stable carbon decreases. The magnitude of this ratio in a sample of groundwater is thus an indication of the length of time since the water left the atmosphere.

Unfortunately, not all the carbon in groundwater comes from the atmosphere or the soil. As we have seen, some can come from dissolution of calcium carbonate, which has probably been part of the aquifer for so long that it contains only stable carbon. Corrections can be made for this, but for these and other reasons isotopic dating (also called **radiometric dating**) of groundwater is never exact and is occasionally completely unreliable. It should be regarded as giving only an approximate guide to the time that water has been in an aquifer.

Carbon-14 has a half-life of about 5600 years, which means that in any sample containing ^{14}C, half of the ^{14}C nuclei will disintegrate in 5600 years. Radiometric dating using ^{14}C is therefore useful for waters that have been underground for some thousands of years, up to a maximum of about 30 000 years. The method has been used to indicate that much of the groundwater present beneath the Sahara desert infiltrated thousands of years ago (page 105).

Another isotope of use in groundwater studies is **tritium**, a heavy isotope (^{3}H) of hydrogen. (Normal hydrogen is ^{1}H.) Tritium, like carbon-14, is produced naturally in the upper atmosphere by the action of cosmic rays on nitrogen. The concentration of tritium is measured in tritium units (TU), one tritium unit being a concentration of one tritium atom in every 10^{18} hydrogen atoms. Under natural conditions the tritium content of precipitation is less than 10 TU. However, like carbon-14, tritium is also formed as a result of the explosion of thermonuclear weapons; between 1952 and 1964 the tritium content of rain was increased considerably, to levels above 2000 TU in parts of the Northern Hemisphere.

The half-life of tritium is 12.4 years, so the presence of tritium concentrations in excess of 5 TU in groundwater implies that the aquifer has received recharge since thermonuclear tests began in 1952.

The measurement of tritium concentrations requires sensitive, expensive equipment. But for this, tritium has been considered by some workers to be the closest thing to the ideal tracer, since it actually forms part of the water molecule.

One of the conditions laid down for an ideal tracer, such as a radio-isotope, is that it should move at the same rate as the water and affected in the same way by chemical and physical processes; it should not, for example, diffuse through shales at different rates or be evaporated at different rates, otherwise it may be subject to unquantifiable concentration or dilution. Such changes in concentration are termed **fractionation**. For a radio-isotope such as tritium to be useful, it is necessary that it should not be subject to significant fractionation, and that its concentration should change only as a result of radioactive decay.

There are however other isotopes – stable isotopes – which are useful in hydrological studies precisely because they *are* subject to fractionation and because they are not radioactive and do not change with time. Two isotopes of particular interest to hydrologists are those which can occur in water molecules – a stable isotope of oxygen, ^{18}O (oxygen-18), and a stable isotope of hydrogen, ^{2}H (commonly called deuterium). The proportions of these isotopes to the common stable isotopes, ^{16}O and ^{1}H, can be measured with a mass spectrometer, and expressed relative to their concentration in an arbitrary standard water ('standard mean ocean water' or 'SMOW').

Because of their greater masses, ^{18}O and ^{2}H isotopes are less likely to evaporate and are more likely to condense than are the more usual ^{16}O and ^{1}H isotopes. As water evaporates from the seas and moves inland to begin its journey through the water cycle, many evaporations and condensations may take place. Generally, the further the water moves from the ocean, the lower is its concentration of these heavy isotopes. The fractionation is also temperature-dependent.

Once the precipitation reaches the ground and infiltrates, no further perceptible fractionation occurs unless the water percolates to great depths and reacts with the rock. The ratios of the heavy isotopes to the common isotopes therefore remain unchanged. In the right circumstances, knowledge of the isotopic ratios of groundwater can provide information about the recharge area from which it was derived or – in large basins where waters from several sources become mixed – about the proportions of water from different sources.

When soil water evaporates, the remaining water becomes enriched in ^{2}H and ^{18}O isotopes. The relative enrichment in the two isotopes is different from that which occurs during condensation. Study of isotopic fractionation of soil water is thus a powerful tool for studying soil-water processes, especially in arid regions where evaporation rates are high and fractionation is pronounced.

Groundwater temperature

One of the important physical aspects of groundwater quality is its temperature. Groundwater in the upper few metres of the Earth's crust experiences seasonal fluctuations of temperature. These decrease with depth and generally become negligible, in temperate regions, below 10 m. At that depth the temperature is about equal to the mean annual air temperature (in Britain this is about $10\,^\circ\text{C}$–$12\,^\circ\text{C}$) and is remarkably constant.

At greater depths the temperature stays constant with time but increases with depth. This is mainly because heat is being generated by the decay of radioactive minerals – the Earth is acting as a natural nuclear reactor. The amount of heat being generated and therefore the temperature rise with depth varies from place to place. In general it is least in old stable areas like the Canadian Shield, and greatest in areas of recent tectonic and volcanic activity, such as Iceland. Britain has a temperature rise with depth – the **geothermal temperature gradient** – which is near the world average of about $25\,^\circ\text{C/km}$.

In most areas of the world, groundwater at high temperature is therefore found only where water has circulated to great depths. Hot springs have been known in some places for centuries and have formed the basis of resorts; those at Bath are an example. There, meteoric groundwater derived from recharge on the Carboniferous Limestone of the Mendip Hills is believed to circulate to depths of more than 4 km. It returns to the surface along fault zones, to emerge as the famous springs, with temperatures as high as $46\,^\circ\text{C}$.

The concern for future energy supplies has led to suggestions that the Earth's natural heat (**geothermal energy**) could be exploited. In some areas this is already done. In Iceland, for example, where the temperature of some of the groundwater is so high that it exists as steam, geothermal energy is used to provide space heating for buildings and for horticulture, and for power generation. Effective use is made of geothermal energy in Italy, New Zealand, Japan, the Philippines and parts of the USA – all areas with a history of recent volcanic or tectonic activity. Now schemes are in progress to try to use geothermal energy in more stable areas with lower geothermal gradients.

A problem is that in these areas it is necessary to drill to great depths to encounter high temperatures; at these depths, permeabilities are often so reduced as to limit well yields. Furthermore, the only techniques available for drilling, testing and producing these wells are those of the oil industry. The high cost of these techniques is justified when producing oil; when applied to hot-water production, however, it means that it costs almost as much to produce a barrel of hot water as a barrel of oil. Since the energy content of the water is much lower, geothermal energy is at present only economical in special circumstances.

Nevertheless, geothermal energy is being used for example to heat buildings in and around Paris, and, in Britain, trial drilling for a demonstration programme has taken place at Southampton and on Humberside. Whether these schemes prove to be economical will depend to a large extent on how much other fuel costs increase in future years.

Selected references

Freeze, R. A. and J. A. Cherry 1979. *Groundwater*. Englewood Cliffs, NJ: Prentice-Hall. (See especially Chs 3 & 7.)

Rodda, J. C., R. A. Downing and F. M. Law 1976. *Systematic hydrology*. London: Newnes-Butterworths. (See especially Ch. 7.)

Slade, J. S. 1985. Viruses and drinking water. *J. Inst. Water Engrs and Scientists* **39**, 71–80.

12 Groundwater: friend or foe?

We saw in Chapter 7 that about one-third of the public water supplies of England and Wales are derived from groundwater, although many of the people supplied from groundwater sources are probably unaware of the fact. How does groundwater travel from an aquifer to the tap of a consumer? There are many possible ways; a typical British layout is shown in Figure 12.1.

In this arrangement, water is pumped from the well at the **pumping station** to a large tank or reservoir. Not to be confused with a large surface-water storage reservoir, this **distribution reservoir** or **service reservoir** will usually be sited on the highest land in the area which it is to serve. From this high point, water has sufficient head to flow by gravity through the distribution mains to the surrounding towns or villages.

Distribution reservoirs usually appear as rectangular humps surmounting hills, though they are often made as inconspicuous as possible. In flat areas the more conspicuous water towers are a familiar sight; they

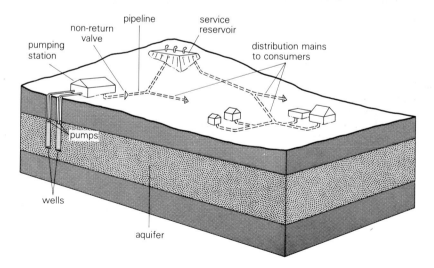

Figure 12.1 A water-distribution system in an area supplied from groundwater sources

serve the same purpose of providing a head of water for local distribution, and a constant head for the borehole pump to work against.

Distribution reservoirs provide a small element of storage, each usually holding a volume of water equivalent to about a day's normal supply for the area which it serves. This means that if a pump fails, consumers will still receive water, and there is an adequate reserve to cope with demand at peak times and for emergencies like fire-fighting. This way the well pump can be smaller than it would need to be if it had to cope with all eventualities.

Water engineers are prudent people. Not only do they nearly always have more wells and pumps serving each area than are strictly necessary, they usually arrange that the operating pump can supply all the water necessary by pumping for less than 24 hours each day. Sometimes the pump operates 'on demand', being switched on automatically when the level in the reservoir falls below a pre-set limit, and switched off when the reservoir is full. More often this arrangement is combined with a time switch, so that as much pumping as possible takes place at preferred times (often at night, to use cheap electricity).

Conjunctive use

In areas underlain by an aquifer, the system outlined above is very convenient. Wells can be sited in or near the areas they have to serve, avoiding the need for long and costly pipelines. If demand increases, new wells can be sunk, so the system can keep pace with the demand. But not all areas with a demand for water are underlain by aquifers – or, if they are, the aquifers may already be fully exploited. In these cases, how can water engineers meet the increased demand for water that results from population growth, from increased industrialisation, or simply from the greater consumption of water that goes with an improved standard of living?

London provides a good example. The aquifers beneath London (principally the Chalk) are fully developed. Similarly, the natural flow of the Thames is already exploited to the full in dry weather. Large areas of land, especially to the west of London, have been used to create storage reservoirs which can be replenished from the river during high flows; it would be difficult and expensive to build more reservoirs.

One possibility is to use the storage space that has been created in the aquifers below London as a result of the lowering of the potentiometric surface; suitably treated water could be pumped into the aquifer during times of surplus river flow, and pumped out again as necessary. Trials of this technique, called **artificial recharge,** have been carried out in the valley of the River Lea, a tributary of the Thames.

Another possibility is to bring water from elsewhere. West of London, beneath the Berkshire Downs, the Chalk contains enormous reserves of usable groundwater; other reserves exist in the Jurassic limestones of the Cotswolds. It would be possible to pump water from these aquifers and convey it to London by a pipeline; however, a natural 'pipeline' already exists in the form of the River Thames. A scheme has therefore been developed that involves pumping water from the aquifers into the Thames to supplement the dry-weather flow of the river. The additional water can then be abstracted downstream using existing river intakes and treatment plants.

This **river augmentation** procedure has several advantages over using pipelines to convey water. To begin with, the cost of the pipeline is saved. Further, not only is the water available at the final destination: on the way, it is adding to the flow of the river for navigation, recreation, fisheries and sewage dilution – unlike water conveyed by pipeline, which cannot serve any of these functions. A disadvantage is that clean groundwater is mixed with river water and will therefore require more treatment before it can be used for public supply than would have been required had it travelled through a pipeline.

In 1984, river-augmentation schemes were in operation or in various stages of development in thirteen British river catchments, including the Yorkshire Ouse, the Severn, the Great Ouse and the Itchen, as well as the Thames. To work successfully these schemes require careful investigation and planning, particularly in the siting of the production wells. If these wells are too close to the river or its tributaries they simply intercept groundwater that would probably have flowed to the river anyway, so the **net gain** (the additional contribution to river flow) is negligible. River-augmentation schemes are usually developed slowly, with observations, measurements, models and then preliminary schemes which are operated for several years.

These schemes are examples of the **conjunctive use** of groundwater and surface water. One feature common to most conjuctive-use schemes is that they use groundwater storage to augment surface-water supplies in dry weather. This usually results in the water table being lowered below its normal minimum level, but this is not a disadvantage: the extra volume dewatered represents storage that would otherwise not be used (Fig. 12.2). This storage is usually quickly replenished in winter, the only difference being that it takes longer for the water table to return to its normal maximum and therefore longer for all the springs and streams to

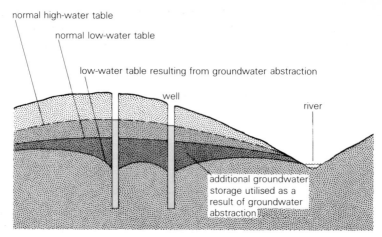

Figure 12.2 Extra groundwater storage utilised in a conjunctive-use scheme

begin flowing again. This is an advantage: as there is generally a surplus of water in winter, this flow is really a waste of water and may even be a nuisance. The reduced winter flow should not be seen, as it sometimes is, as the first stage in the 'drying up' of the aquifer.

Groundwater as a problem

It is convenient when we require a water supply to be able to drill a well into an aquifer, insert a pump, and abstract water. What is less convenient is that groundwater will flow into any excavation in the saturated zone, unless we somehow prevent it. Whether it is a mine shaft, a gravel pit, a railway tunnel, or simply the basement of a house, if it is within the saturated zone it is going to be affected by groundwater.

The effects do not consist merely of the entry of water; the presence of groundwater can also affect the strength and stability of the ground. Anyone who has tried to dig a hole in a sandy beach will have experienced both of these effects. When a hole is dug into firm damp sand on a beach, the sides of the hole, to begin with, stand without being supported. As soon as the hole reaches the level of the sea or lake, however, we strike water – we have in effect reached the local water table. At that point, usually, water and sand flow in together, and the bottom of our hole turns into a wet sandy mess. There in miniature are the two problems that groundwater can cause in excavations – flooding and instability.

The extent to which the inflow of groundwater causes a problem is dependent on its quantity, and hence on the permeability of the surrounding rock or soil. The extent to which the groundwater causes instability however is dependent on the pore-water pressure, and need bear no relationship to quantity or permeability; indeed, it is often more of a problem in poorly permeable material than in highly permeable rocks.

Groundwater as a cause of instability

We can return to the inflow problem in a while. First let us consider the question of instability.

We saw in Figure 7.3 that the weight of a confining bed is supported partly by the aquifer framework and partly by the water in the pore space of the aquifer. Actually, these considerations apply to any level of any porous rock unit. Suppose we consider the forces acting on a horizontal plane at some arbitrary depth (Fig. 12.3). The pressure acting downwards on this plane is the weight of the overlying material divided by the area of the plane. This is balanced by the upward pressures on the plane, which are the grain-to-grain contact pressure w, and the pore-water pressure p (above atmospheric), i.e.

$$W = w + p \qquad (12.1)$$

The study of these and similar forces within consolidated rock is termed 'rock mechanics'; within unconsolidated material it is called **soil mechanics.** In soil-mechanics terms, the total pressure W is called the

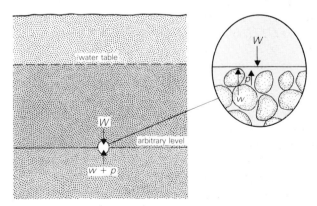

Figure 12.3 Total stress and effective stress At any level in the ground, the pressure W (the total stress) exerted by the overlying material is balanced by the average grain contact pressure w (the effective stress), and the pore-water pressure, p. (p is measured relative to atmospheric pressure.)

total stress; the portion w that is supported by grain-to-grain contact is termed the **effective stress**. In many situations in hydrogeology the total stress W does not change, because the weight of the overlying material is unaltered. What *can* change however are the relative sizes of w and p, i.e. the proportions of the total stress carried by the framework and the water. As we saw in Chapter 7, if p is reduced then w must increase, and the aquifer may be compacted as a result of the increased effective stress, leading to subsidence.

If the total stress W is reduced for some reason – perhaps by excavation of some of the overlying rock or soil – or if p is increased, then the effective stress w will be reduced. This means that the grains will not be pressed so hard against each other. If W were reduced or p increased to the point where $W = p$, then all the weight of the overlying material would be borne by the pore water; the effective stress would be reduced to zero, because there would no longer be any significant grain-to-grain pressure.

In consolidated rock such as sandstone the grains are held together by cement. In most unconsolidated materials it is the friction between grains resulting from the effective stress that gives the material most of its **shear strength** – its resistance to movement of the particles across each other. In the absence of any effective stress, there is a tendency for the particles of most sediments to stick together slightly, particularly if clay minerals are present. This gives rise to some shear strength, called **cohesion**, even when the effective stress is zero.

Clean sand has little or no cohesion; if the pore pressure rises or the total stress falls so that the effective stress is reduced to zero, the sand loses its shear strength and behaves more or less as a liquid; this is the **quicksand** condition, much quoted by writers of adventure stories.

Even if the effective stress is not reduced to zero, the reduction in shear strength caused by a decrease in W or an increase in pore pressure can sometimes lead to material slumping into excavations (as into our hole on the beach) or to the occurrence of landslips. Many landslips occur during or at the end of the rainy season, or after exceptional rainfall. Unconsolidated sand or silt which flows into excavations or boreholes because of the reduction in effective stress is often referred to as 'running sand', as though it were a particular type of material.

An interesting question is: what happens when the pore-water pressure is less than atmospheric, i.e. when p in Equation 12.1 is negative? In other words, what happens in the unsaturated zone? Use of Equation 12.1 suggests that if p is negative (as it is above the water table) then w is greater than W, i.e. the effective stress is greater than the total stress and the water is helping to hold the grains in contact. It might seem reasonable to suppose that the surface-tension forces in the unsaturated zone do help to hold the grains in contact, just as the water films around sand grains mean that wet sand will cling to a knife blade whereas dry sand will not. The sand in our hole on the beach was stable above the water table and only became unstable where it was fully saturated. However, the stress relations in the unsaturated zone are more complicated than these simple examples might suggest, and in general Equation 12.1 should not be relied upon in the unsaturated zone.

In consolidated rock, movement along joints and faults can be influenced by changes in pore pressure. A particular example is the occurrence of earthquakes. Stresses associated with the movement of crustal plates build up in the rocks at various places in the Earth's crust. The stresses increase until they exceed the strength of the rock, which abruptly fractures. The resulting fracture is called a **fault**; the sudden release of energy causes the **earthquake**. Once a fault has developed – usually at a weak place in the rock – subsequent movements tend to occur along it; stresses build up until they overcome the resistance to movement caused by the friction of the rocks along the fault surface. When the frictional forces – which depend on the effective stress and hence on the pore pressure – are exceeded, an earthquake occurs. This suggests the possibility that we may be able to predict earthquakes by monitoring the fluid pressures along the major fault zones where earthquakes occur. Even better, by injecting water into these zones we may someday be able to control earthquakes. By keeping frictional resistance low, we might be able to prevent large stresses from building up; in this way we could ensure that the stresses were relieved a little at a time, without the dramatic releases of energy that accompany major movements.

In general, studies of pore-water pressure problems and the measures used to combat them are an essential part of the work of the engineering geologist, and reference should be made to textbooks on that subject for a full treatment of them. Suffice to say here that there are few problems of soil and rock mechanics that are not in some way influenced by groundwater, and it is unfortunate that engineering geologists and hydrogeologists often use different terminology to describe the same thing.

One particular example of pore-water pressure as a problem occurs in connection with concrete dams. By its very purpose, a dam causes a considerable difference in head between its upstream and downstream sides. Concrete dams should always be sited on rocks of low permeability but, as we saw in Chapter 6, no rock is completely impermeable and under the influence of the head difference water will flow under the dam. This flow is undesirable not only because of the loss of water, but because it

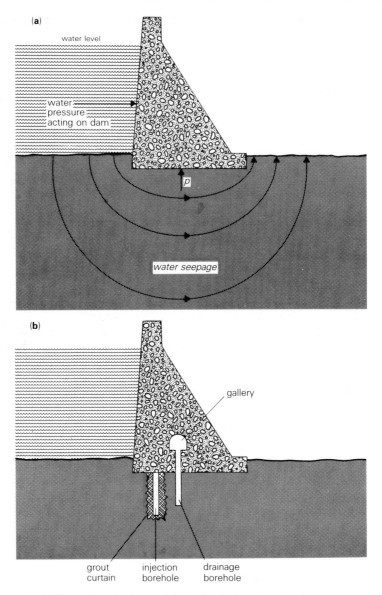

Figure 12.4 Seepage under dams (a) The head of water behind a concrete gravity dam causes increased pore-water pressure beneath the dam; this in turn decreases the frictional resistance that is preventing the dam from being pushed downstream by the water behind it. (b) A grout curtain reduces the seepage of water beneath the dam, and drainage boreholes relieve the pore-water pressure. A diaphragm wall may be used instead of a grout curtain.

leads to increased pore pressures in the rock beneath. The resulting reduction in effective stress reduces the frictional forces that are preventing the dam from being moved downstream by the pressure of the impounded water (Fig. 12.4a).

The seepage of water beneath the dam tends to decrease with depth. To lessen the effects, engineers use two techniques. First, they inject grout (pp. 114 and 178) into the rock from a line of boreholes drilled beneath the dam. The grout strengthens the rock and reduces its permeability and porosity, leading to a reduction in seepage. Second, they drill another line of boreholes parallel to the grout curtain. These boreholes relieve any build-up of pore pressure by allowing the water to drain into a gallery within the dam, from where it can be removed (Fig. 12.4b).

Groundwater inflow to excavations

The inflow of groundwater into excavations can be prevented or reduced in a variety of ways. The technique chosen will depend on factors like the permeability and the hydraulic gradients present, whether it is also necessary to stabilise the ground, the size of the excavation, and whether control is to be temporary or permanent. Sometimes one technique is used as a temporary control while more permanent controls are being installed; frequently two or more techniques are used in combination. The more common techniques are listed below.

Sheet piling
This is a groundwater-exclusion technique, used where saturated, permeable unconsolidated materials overlie relatively impermeable materials at shallow depth. Corrugated steel sheets are driven down to the impermeable layer from the surface. The method provides reasonable ground support at the sides of excavations, but rarely provides a complete barrier to groundwater movement.

Diaphragm walling
This functions in a similar way to sheet piling, but sophisticated emplacement techniques are used to install a thin concrete wall (the diaphragm wall) through the permeable material. It is more expensive than sheet piling but provides a much more effective barrier to groundwater movement.

Grouting

In simple terms grouting involves injecting into the pore spaces of the rock a liquid (called **grout**) that will harden and set. The method is used to permanently reduce permeability in all rock types and − especially in unconsolidated rocks − to help stabilise the ground. Grout is injected through specially drilled boreholes, and packers (p. 147) may be used to isolate certain intervals for grout treatment. The two main types of grout are **cement grouts** (which are essentially thin concrete, sometimes containing clay) and **chemical grouts**. The cement grouts are sometimes called **particulate grouts** because they consist essentially of particles suspended in water; because of this there is a lower limit of about 0.2 mm to the size of pore or fissure into which they can be injected, because of the natural filtering action of the rock. Chemical grouts are similar to epoxy-resin adhesives; they do not suffer to the same extent from the filtering problem, and depending on their viscosity may penetrate for considerable distances into the formation. Large volumes of expensive thin chemical grout can be 'lost' along fissures or into permeable layers unless care is taken. The usual procedure is to grout with a cement grout first to seal major fissures or very permeable layers; when this has hardened, successive applications with chemical grouts of reducing viscosity are used to seal the less permeable intervening material. It is impossible, however, to fill completely all the pore space in a rock.

It is important that grouting should be carried out before excavation, because the presence of the excavation will usually cause a hydraulic gradient towards it which may make it difficult to place the grout where it is needed. Detailed investigations, particularly of permeability distribution, are necessary before any grouting operation if costs are to be predicted accurately.

Freezing

The technique of freezing to provide a temporary reduction in permeability dates back to the end of the 19th century. It is most frequently used in association with the sinking of mine shafts. Boreholes, cased throughout their depth, are drilled around the shaft site, and refrigerated brine is pumped into each borehole through an access pipe. The brine is kept in circulation, cooling the rock and freezing the groundwater. A cylinder of frozen ground thus forms around each borehole and these solid cylinders eventually combine to surround the shaft site with an ice barrier. The excavation of the shaft can then proceed within the protection of the frozen ground. Once the shaft is completed and lined, the ground is allowed to thaw.

Freezing is expensive, but the costs can be predicted with reasonable accuracy; grouting costs on the other hand can escalate if the permeability variations are not well known in advance. Freezing will usually provide the only practical means of dealing with large thicknesses of saturated fine-grained materials, which are difficult or impossible to grout.

Compressed air
In tunnelling operations, the entry of groundwater into the unlined portion of the tunnel is sometimes reduced by artificially increasing the air pressure – typically to about twice normal atmospheric pressure – until the permanent waterproof lining can be installed. The increased pressure reduces the head difference between the rock and the tunnel, and therefore reduces the inflow.

This technique means that the workmen must work in an artificially high air pressure, entering and leaving through an air lock. If the pressure is significantly above atmospheric strict precautions have to be taken, including provision of depressurisation facilities similar to those used by deep-sea divers.

Groundwater lowering
This technique, also known as **dewatering,** involves abstracting groundwater so as to lower the water table and permit excavation or other work to proceed within the dewatered area (Fig. 12.5). The dewatering may be temporary – perhaps for the construction of a shallow road or railway tunnel which will eventually have a waterproof lining – or permanent. Permanent lowering of the water table is used so that buildings can be constructed in areas such as river flood plains, where the water table is close to ground level.

In confined formations groundwater may be abstracted to reduce the head and hence the inflows into underground openings such as shafts. The technique can be particularly effective if used in association with grouting.

In theory, dewatering can be achieved in several ways. A common technique for temporary dewatering is to install rows of small-diameter boreholes called **well points** around the site. These are usually driven or washed into place by pumping water into them as they are forced into the ground – a process called **jetting.** When they are in place, several are connected to a common suction pipe and pumped using a surface pump. Because of the lift limitations of this type of pump, successive lines of well points may have to be installed at different levels within the excavation if the water table is to be lowered more than a few metres.

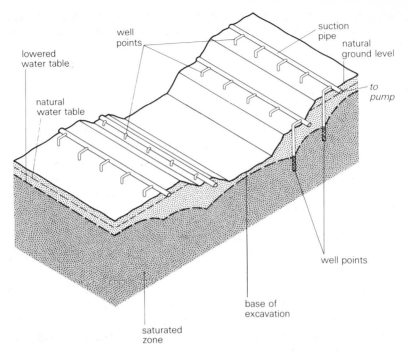

Figure 12.5   Site dewatering   A two-stage dewatering scheme, using well points to lower the water table below the base of an excavation.

Drainage

Whatever method is used to reduce the permeability of the rock around an excavation, there is still usually some flow of water into the excavation which has to be pumped out. The usual way of doing this is to drain the water to sumps at the lowest part of the excavation, and to pump it out from there.

If the excavation is in rock of low permeability, then drainage of this kind may be the simplest and cheapest way of dealing with groundwater, without resort to other techniques. It may even be an adequate long-term solution, for example where road or railway cuttings pass just below the water table for part of their length.

Sometimes drainage operations can unintentionally become dewatering schemes, with surprising results. London Regional Transport's District Line passes through cuttings or 'cut-and-cover' tunnels for much of its length. The drainage of the railway has exerted a considerable influence on the groundwater-flow patterns in the gravels north of the Thames, the full effect only being discovered during the course of investigations for the Thames Barrier. Between West Kensington and Temple stations, 6800 m³ of water is pumped out each day to keep the track dry.

Mine drainage

Deep mines inevitably have most of their workings below the water table. It would not be practical or economical to line all of the workings to exclude groundwater, but fortunately most mines are in rocks of low permeability and the workings are kept dry by pumping. Metalliferous ores are usually mined from metamorphic rocks. At depth these rocks rarely have open joints or fractures, so that seepage into the workings is slow except in the vicinity of major faults or zones of shattered rock, which can generally be grouted or lined individually.

In Britain, the Carboniferous rocks in which coal occurs are of relatively low permeability, so that it is usually possible to pump water from the workings faster than it can enter. The Coal Measures are effectively separated into hydraulically distinct units by beds of clay. Mine drainage may locally dewater some of these units − depressing the potentiometric surface for that unit by hundreds of metres − while strata above and below remain saturated.

In some coalfields the Carboniferous rocks are overlain by saturated permeable Permo-Triassic sandstones. These sandstones are separated from the workings by less permeable Carboniferous strata, but the access shafts to the mines must penetrate the permeable material. Several such shafts have recently been sunk for the new mine near Selby in Yorkshire, where freezing, grouting and pumping to reduce the groundwater head were used to permit excavation. On completion, shafts like these are lined to exclude water, since pumping alone could not keep them dry. Early shafts were lined with brick or with cast-iron linings called 'tubbing', but modern shafts are generally lined with concrete.

The water pumped from mine workings frequently has a high content of dissolved solids (Ch. 11), including iron compounds. Its disposal may therefore require some care: if merely pumped into surface drainage or onto permeable ground it could contaminate river water or local aquifers, and cause unsightly iron-oxide deposits.

When workings are abandoned, it may be necessary to continue pumping to protect adjacent workings. If pumping from a disused mine is stopped, the mine will become filled with water, and may endanger later underground operations in the area. A terrible example of this occurred in March 1973 at Lofthouse Colliery, Yorkshire, when a new coalface was excavated too close to old flooded workings, abandoned in the last century and unknown to the modern mine's surveyors. In the sudden inrush of water, seven men were killed. The report of the Public Inquiry is a grim warning to make the fullest study of historical records before commencing new works.

Selected references

Freeze, R. A. and J. A. Cherry 1979. *Groundwater.* Englewood Cliffs, NJ: Prentice-Hall. (See especially Ch. 10.)

Gray, D. A. and S. S. D. Foster 1972. Urban influences upon groundwater conditions in Thames Flood Plain deposits of Central London. *Phil. Trans. R. Soc. Lond. A* **272**, 245–57.

Hardcastle, B. J. 1978. From concept to commissioning. In *Proceedings of conference, Thames Groundwater Scheme,* 5–31. London: Institution of Civil Engineers.

Mines Inspectorate 1973. *Inrush at Lofthouse Colliery, Yorkshire.* Cmnd 5419. London: HMSO.

Todd, D. K. 1980. *Groundwater hydrology,* 2nd edn. New York: Wiley.

13   Some current problems

The science of hydrogeology developed because people needed to get
water supplies from the ground. To begin with, in the Old World, little
science was needed; the benefit of experience, passed from one generation to the next, was available to tell people where to dig and where not
to dig wells. But as industry and agriculture began to demand ever larger
supplies, local experience was not always adequate; and as demand arose
in newly settled areas like the American Mid-West and the African col-
onies, the experience was often non-existent. Thus arose the need for
investigation and scientific guidance, leading to the work of people like
Darton and Meinzer of the US Geological Survey, and of Frank Dixey
in the Colonial Surveys in Africa.

To begin with, the emphasis was often simply on finding suitable
aquifers. Later, the realisation that resources were finite and the evidence
of declining heads in some places (as in the Dakotas) caused more attention to be given to understanding aquifers, measuring their properties
and estimating their resources. Hubbert, and Theis and other scientists
of the USGS, made major contributions during this period.

Initially, provided that the water was drinkable (by either people or
animals, as appropriate), quality was not considered a major factor –
quantity and reliability were more important. Since World War II and
particularly since 1960, groundwater studies have broadened to include
much more than quantitative water-resources investigations. Studies of
groundwater quality have expanded and in particular much emphasis has
been put on studying pollution and the movement of contaminants
within groundwater systems. The pollution may arise accidentally, as for
example when oil or chemicals are spilled as a result of a road or rail acci-
dent; it may be intentional, as when hazardous wastes are disposed of by
burying them in excavations or by injecting liquid wastes into shafts
or boreholes; or it may be semi-intentional or incidental. In this last
category we can include the effects of fertilisers and other chemicals,
which are applied for good agricultural reasons but which may be
leached from the soil and carried down to the water table during
infiltration.

The safe disposal of hazardous wastes in such a way that they do not
cause harmful contamination of aquifers is now an important facet of
hydrogeology. For most wastes the generally accepted philosophy is that
it is advisable to allow them to be gradually dissolved or diluted by

groundwater and flushed away in concentrations which are harmless by the time they reach sources of water supply – this is the so-called 'dilute and disperse' approach. Sometimes very toxic chemical wastes resulting from some industrial processes are injected into deep formations containing saline water, which are not themselves used for any water supply and which are separated from aquifers by beds of relatively impermeable material.

A particular cause of concern in many countries at present is the highly radioactive wastes which result from the use of uranium fuel in nuclear-power stations. It is suggested that a way of disposing of these wastes would be to incorporate them in some relatively inert material such as borosilicate glass and then to bury them deep below ground in an impermeable 'host' rock. Because these materials are so dangerous, and will remain so for many thousands of years until their radioactivity decays, it is felt that the 'dilute and disperse' doctrine is not applicable and that they must be kept isolated from groundwater. It is groundwater that would slowly corrode the metal cannisters holding the wastes, leach the radioactive materials from the borosilicate glass and, possibly, eventually transport them to man's environment. The task of identifying 'impermeable' host rocks where this transport will not occur for at least 100 000 years is currently occupying teams of hydrogeologists and other Earth scientists in many countries. A problem is that many of the rocks that would be regarded as impermeable in a normal hydrogeological investigation have significant permeability when tens of thousands of years are available for movement. Another problem is that the heat given out by the radioactive waste may have a deleterious effect on some potential host rocks such as clay or rock salt. A third problem is that in rocks that are almost impermeable, the limited hydraulic pathways that do exist – such as fissures in granites or metamorphic rocks – are often unevenly distributed, making it difficult to predict *how* flow will take place if it does occur. These studies have led to a major interest in the hydrogeology of fissured rocks.

A phenomenon that plays a major part in pollutant movement is **dispersion**. If a small volume of pollutant (or a chemical tracer representing it) is released instantaneously at a point in an aquifer, that tracer will not retain its fixed volume. As a result of molecular diffusion and more importantly of mechanical dispersion, it will spread out to form a plume both along (longitudinal dispersion) and perpendicular to (transverse dispersion) the flow direction, becoming diluted in the process (Fig. 13.1a). This means that it will not arrive at a downstream sampling point at a single instant, but over a period of time that will increase with distance travelled. The **mechanical dispersion** arises chiefly from the

(a)

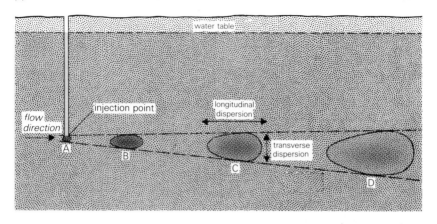

(b)

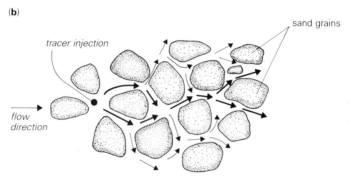

Figure 13.1 Dispersion (a) Dispersion in a homogeneous isotropic aquifer. A fixed volume of tracer is released at the injection point A at time 0. At time t the tracer has reached B; after time t′ it has reached C; after time t″ it has reached D. Note that dilution and dispersion increase with time and distance travelled, and that longitudinal dispersion is greater than transverse dispersion. (b) The process of mechanical dispersion in a sandstone.

separation, recombination and tortuosity of the pore channels (Fig. 13.1b) – or the fissures in a fissured aquifer – and from the different speeds of groundwater flow in channels or fissures of different widths. Attempts have been made to derive general relationships for the dispersion characteristics of aquifers, but it has now been realised that minor variations in permeability and porosity within an aquifer can exert a major influence on the degree to which dispersion occurs at any specific locality.

An example of incidental contamination can occur in coastal aquifers. In a coastal aquifer, fresh water derived from infiltration overlies saline

(a)

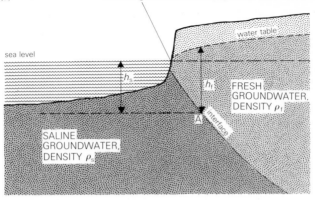

zone of groundwater seepage from land to sea

(b)

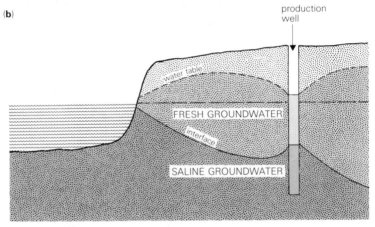

production well

Figure 13.2   Coastal aquifers (a) Under natural conditions there is usually a flow of groundwater from land to sea. The pressure of fresh water ($\varrho_f g h_f$) at A equals or exceeds the pressure of saline water ($\varrho_s g h_s$). (b) When pumping takes place, the lowering of the water table induces a corresponding rise of the interface; saline water migrates inland and may eventually reach the well.

Static head and water pressure (Fig. 13.2)

Because of the lower density of fresh water, longer columns of fresh water than salt water are needed to give the same pressure. Strictly, we should be talking about static heads, not pressures, and we should express all the heads in terms of columns of one liquid of fixed density. However, at any point on the interface the elevation head is fixed and the static head is therefore directly proportional to the pressure. It is therefore acceptable to talk in terms of pressure, and in this case it is probably an easier concept to work with.

Figure 13.3   Water problems in the Third World   Children in a Bangladesh village take advantage of water flowing in an irrigation channel from a deep borehole. Despite the efforts of the Bangladesh Government and the aid agencies, many villagers still have to walk hundreds of metres to obtain drinking water.

water (Fig. 13.2a) in such a way that at the interface between them the pressure of fresh water usually exceeds the pressure of denser salt water, causing flow to occur from land to sea.

If the natural situation is disturbed by pumping, the lowering of the water table results in a corresponding upconing of the interface, and salt water may be drawn into the well (Fig. 13.2b). In practice the interface is not sharp but is affected by dispersion. Salt-water intrusion can also occur inland, particularly in arid regions, where fresh water overlies saline water. It is a problem that is receiving attention in many parts of the world, including Florida, the Netherlands and Israel.

Hydrogeologists are also now involved in exploring for, and understanding the origins of, hydrocarbon accumulations. Groundwater movement – albeit at greater depths and slower flow rates than those which normally concern hydrogeologists – must have been responsible for the transfer of oil and gas from the source rocks where they formed to the reservoir rocks where they are found and exploited. It follows that a knowledge of the flow conditions can help in locating oil and gas occurrences.

For many years the study of groundwater hydraulics developed parallel to but separate from the work done by petroleum engineers studying the behaviour of oil and gas reservoirs. Now there is an increasing interchange of ideas between the two groups. Exploration for and exploitation of geothermal energy (p. 168), another field in which hydrogeological expertise is needed, offers great scope for collaboration between the water and petroleum sciences.

All these, and others yet more exotic and further removed from water resources, are legitimate fields of interest for hydrogeologists. But the problems that these fields generate and the challenges that they provide tend to pale into insignificance in comparison with the task of providing drinking and irrigation water for the peoples of the Third World (Fig. 13.3). Their problems will be solved not by the increasing sophistication of the specialised few, but by greater awareness of their problems among the many. The main aim of this book is to heighten that awareness.

The Earth has sufficient water for all its population. The advanced nations have the knowledge and ability to solve the problems. It remains to be seen whether they have the wisdom and the will.

Selected references

Freeze, R. A. and J. A. Cherry 1979. *Groundwater*. Englewood Cliffs, NJ: Prentice-Hall. (See especially Ch. 9.)

Todd, D. K 1980. *Groundwater hydrology*, 2nd edn. New York: Wiley. (See especially Chs 8 & 14.)

Index

abstraction 126–7
acid rain 157
aerial photographs 128–9
alluvial aquifers 84
amoebae 154
angle of contact 20–2, 23, *4.1*
anisotropy **49**, *6.9*
aquiclude **85**
aquifers **10**, 12, 34, 50, 55–7, **65**, 66–87, 132–51, *2.3*; alluvial 84; artesian 66–8, 70, 89, *92*; coastal 88, 185, *13.2*; confined **26**, 58, 60, 66, 67–8, 67–74, 76, 138, 139, 141, *2.3*, *6.10*; different rock types 77; discharge from 74, 88, 92–3, *8.3*, *8.4*, *8.5*; elasticity of 69–70, 76; flow through 53–4, 57–60, 63–4, 69, 89, *6.10*; in Denmark 79; in Sweden 79; in United Kingdom 79–84, *7.6*; perched 66, *7.1*; pumping of 139–42; saturated thickness of 54, 57, 60, 63, 140–2, *6.10*; thickness of 54, 55, 116, 132; transmissivity of 54, 55, 60; unconfined **11**, 54, 57, 63, **66**, 68, 69–74, 76, 92, 139–42, *6.10*; water-table **66**
aquifer loss 60, 64, 117, *6.13*
aquifer model *see* model
aquifer properties 132–51
aquitard 85
Aristotle 17
artesian aquifer 66–8, 70, 89, 92
artesian well 55, 66, 68, 72, 108, *7.2*
Artesium 66
artificial recharge 171
Artois 66, 108
atmosphere 8, 13
atmospheric pressure effects on wells 72–3, *7.4*
atomic mass 164
atomic number 164

bacteria 154–5, 158
  coliform 155
bailer 118, 123
Bangladesh 84, 109, *13.3*
barometric efficiency **73**
barometric fluctuation **72**, 74, *7.4*
barrier boundary 132, 139, 142, *10.3*
baseflow **93**, 94, 98–9, 128
baseflow–separation curve **98**, 99
Bath springs 167
biological quality of water 156
borehole 3, 9, 53, 54–9, **136**, 109, 129, 131–2
borehole logging 130–1, 145, *10.1*, *10.2*, *10.7*

borehole television 145, *7.7*
Boulton, N. S. 142
boundary
  barrier 132, 139, 143, *10.3*; hydraulic 132, 138, 139, 144, *10.3*; recharge 132, 139, *10.3*
bourne **94**
British Geological Survey 128
Broad Street pump 154–5

cable tool drilling *see* percussion drilling
calcium bicarbonate 19
calcium carbonate 19–20, 29, 79, 157–8
  dissolution of 19, 81, 157–8
California 76, 84
capillarity 22, 23, 25, *4.2*
capillary (tube) 22–5, 28, 31, 61, *4.2*
capillary fringe **25–8**, 142, *4.3*, *4.4*
capillary water 25
carbon dioxide 19, 29, 81, 157, 165
carbon isotopes 165
Carlsbad Cavern 19
casing 114–15, 117, 118–23
catchment (area) 88–9, 98, 99, 127–8, *8.1*
  groundwater 88–9, 99, *8.2*; influence of rock types 99
caverns 3–6, 19–20
cementation of sediments 78, 81, 84
  of Chalk 79–81; of Permo-Triassic sandstones 83
chalk, Chalk 3–6, 25, 54, 55, 62, **79–82**, 170–1, *2.1*, *7.6*, *7.7*
  catchments 88–9, 99–101, *8.11*; lowering of potentiometric surface 67; transmissivity 54, 55, 81–2; yield–depression curve *6.14b*
channel 35
Chebotarev sequence 160
chemical quality of water 156–63, *11.1*, *11.2*
chisel 118, *9.5*
chlorine 155
cholera 153–4
clay 25, 29, 78, 136, 174
Coal Measures 181
coastal aquifers 88, 185, *13.2*
cohesion 174
coliform bacteria 155
compaction 78, 84, 141
compressed air (control of groundwater) 179
compressibility effects 69, 70
computers 142, 150–1
cone of depression **58**, 59, 64, 116–17, 133, 136–43, *6.11*, *6.12*, *6.13*, *10.6*
  development of *6.11*

confined aquifer **11**, 58, 60, **66**, 67–8, 69–74, 76, 138, 139–41, *2.3*, *6.10*
  pumping of 139
confining bed, layer 11, 66, 68, 72, 85, 92, 133, 138, *2.3*, *7.2*, *8.5*
conjunctive use 171, *12.2*
connate water 156–7, 162–3
consolidated rocks, sediments 78–9, 173–5
  drilling in 118–25; well construction in 108, 113, 114, 117, *9.4*
contaminants 183
contamination 9, 185
continuum 50
cores, coring 123–5, 132, 135, *9.7*
core barrel 124–5, *9.7*
current meter 95–6

Dakota Sandstone 68, 70
dam, flow beneath 50, 175–7, *12.4*
Darcy, H. 45–9, 63, 143
Darcy's law 45, **48**, 50–4, 55–7, 60–2, 63–4, 127, 135, *6.8*, *6.14*
  analogy with Ohm's law 48; experimental verification *6.8*; validity 62
darcy (unit) 51
datum (level) **37**–42, 47, 63, 95
degree of saturation 30
delayed yield 142, *10.6*
deserts 104–5, *8.12*
deuterium 166
development (of wells) 116
dewatering 179, *12.5*
diaphragm wall 177
Dijon 46
discharge of groundwater from aquifers 74, 88, 92–3, *8.3*, *8.4*, *8.5*
discharge of river or stream 95–7, 98–9, *8.8*, *8.9*
discharge area 68, 89, 162, *8.1*, *11.2*
discharge hydrograph 95, 97, 99, *8.9*, *8.10*, *8.11*
diseases, water-borne 153–6
dispersion 184–6, *13.1*
dissolution 157–8, 162, *11.1*, *11.2*
  of calcium carbonate 19–20, 81, 158
distribution reservoir 169–70
distribution system 169
divide 88–9, 127, *8.1*, *8.2*
Downing, R. A. 89
dowser 3, 151
dowsing 151
drainage
  from excavations 179–80; from mines 162, 181; of pore space under gravity 27–8, 30–1, 34, 69, 83–4, 141, 142, *10.6*
drawdown **60**, 62, 64, 111, 117, 138, 139–42, *6.13*
drill bit 118–21, 122, 123, *9.5*, *9.6*
drill cuttings 121, 123–4, *9.6*
drill pipe 119–21, 125, *9.6*

drilling 108, 118–25, *9.5*, *9.6*
  air 122; downhole motor 122, 123; hammer 122; percussion 118, 122; percussion rig 118–21, 123, *9.5*; reverse circulation 121; rotary 118–25; rotary rig 120–2, 123–5, *9.6*
drilling fluid, mud 120–4, 132, *9.6*, *9.7*
drilling string 118–22
drought 101, 103–4
  of 1975–76 101–4
Dupuit–Thiem equation 138
dynamic viscosity 51
dysentery 154

earthquakes 175
  cause of water-level fluctuations 74
efficiency of wells 60, 116
elasticity of aquifers 69–70, 76
electrical resistivity surveying 130–1
electromagnetic radiation 129
elevation energy **39**, 40–3, 44, *6.3*, *6.5*, *6.7*
energy 37–45, 59–60, 63, *6.5*
  law of energy conservation 40, 42–3
engineering geology 175
ephemeral stream **93**, 104
Escherichia coli 155
Euphrates, River 92
evaporation 13–15, 16, 17, 30–1, *3.1*
  measurement of 33
evapotranspiration **15**, 32–4, 75, 99, 102–3, 126–7, 128
excavations 172, 174, 177–8

faults 132, 143, **175**, 181, *8.4*, *10.3*
fertilisers, leachate from 158, 183
field capacity **30**, 31–2, 34, 35, 102
field measurement of aquifer properties 134, 136–48
filter cake 121, 123
filter pack **114**, **116**, 123, *9.4*
fissure 20, 55, 144, 145–8, 184–5, *2.1*, *7.7*, *10.7*
  detecting with logging 145; drilling problems 122–3; in chalk 81–2, *7.7*; in Permo-Triassic sandstones 83
fissure flow *10.7*
flow, laminar 60–2
  of rivers and streams 93–4, 98, 99–101, 103, 126–7; analysis of 97–101; measurement of 94–7; overland **15**, 35, 88, 98, *3.1*; steady 42, *6.6*; through an aquifer 53–4, 57–60, 63–4, 68, 89, *6.10*; through a constriction *6.7*; through a pipe 42–6, *6.6*, *6.7*; through rock 45; through sand and sandstone 44–7, *6.9*; to excavations 172, 177–80; to a well 54, 55–60, 64, 117, 138, *6.11*, *6.12*; turbulent 60–2; vertical components of 54, 58, 63, 66, 69, 85, 89, 142, 143, *6.10*, *8.1*
flowmeter, borehole 145, *10.7*

flume 96–7
fossil water 105
Fourier's law 48
fractionation, isotopic 166–7
freezing to control groundwater 178, 181
friction 42–3, 45

gaining stream **92**, 93, 101, 103, 133
Ganges delta 84
gauging station 95, 96, 99, 127, *8.7, 8.8*
geological map 128
of United Kingdom *7.6*
geological measurements 126, 128
geophysical exploration 129
geophysical logging 130–2, 145, *10.1,
10.2, 10.7*
geothermal energy 168, 187
geothermal temperature gradient 167
groundwater **8**, 9–12, 18
advantages of 7–9; age of 105, 164–5;
cause of instability 172, 173–7;
disadvantages of 9; flow paths 88, *8.1*;
flow to excavations 172, 177–81; flow
to rivers 92, 94, 97–101, 127; in
Denmark 79; in England and Wales 94,
169; in Sweden 79; in United Kingdom
79–84; movement of 20, 37, 55, 62–3;
occurrence of 5–8, 16; problems caused
by 172–82; replenishment of 12, 15;
role in hydrocarbon accumulations 187;
treatment of 155
groundwater abstraction 126–7
groundwater catchment 88–9, 99, *8.2*
groundwater chemistry 156–63, *11.1, 11.2*
groundwater discharge *see* aquifers
groundwater lowering 179
groundwater mining 105–7, 163
groundwater quality 105, 145, 153–68,
183
groundwater resources 126
groundwater temperature 156, 167–8
groundwater withdrawal 128
grout 114, 177, **178**, 181, *9.4, 12.4*
gullies 35

Halley, Edmund 17–18
Hantush, M. S. 139
Hard, H. 68–70
hardness 158
hazardous wastes 183–4
head 37–8, **39**, 43–4, 54–5, 63, 72–4,
114, 137, 187, *6.1, 6.3, 6.4, 8.1*
dynamic **42**, 46, 57; elevation 39, 40,
41–2, *6.4, 6.8*; pressure 39, 40–1, *6.2,
6.4, 6.8*; static 44, 57, 66, *6.4*; total 44;
velocity *see* dynamic
head difference **39**, 55, 62, 63, 89, *6.1,
6.7*
head gradient **44**
head loss **44**, 47, 59–60, 62–3, 116, *6.6,
6.7, 6.8*
hepatitis 154
heterogeneous, heterogeneity 49, 50, *6.9*

Hewlett, J. D. 98
homogeneous, homogeneity **49**, 50, 55,
63, 88–9, 138
Horton, R. E. 15, 98
Hortonian flow **15**, 35
house water supply 39, *6.2*
Hubbert, M. K. 52, 183
hydraulic boundary 132, 138, 139, 144,
10.3
hydraulic conductivity 48–9, **50, 51**,
53–4, 55, 63, 132, 135
of chalk 81–2; of igneous and
metamorphic rocks 86; of Permo-
Triassic sandstones 83
hydraulic gradient **44**, 47–8, 50, 53–5,
57–9, 63, 127, 133, 138, *6.8*
in chalk 82–3; steepening near well
6.12; vertical 54–5, 58, 60, 68, 89, 143,
6.10, 8.1
hydraulic head *see* head
hydraulic measurements 126, 132–48
hydrocarbon accumulations 187
hydrogen isotopes 165–7
hydrogeological map 128
hydrogeology **2**, 18, 126
development of 183
hydrograph 94–101
discharge 94, 97, 99, *8.9, 8.10, 8.11*;
stage 95
hydrograph analysis 98–101, 127, *8.9*
hydrological cycle 16, 18, 126, 156–7, *3.1*
hydrological measurements 126
hydrology **16**, 18

igneous intrusion 77
igneous rock **77**, 79, 86
extrusive 77, 86; intrusive 86
impeller **109**, 111, *9.1, 9.2*
impermeable 10, 50, 85
India 84
Indus Valley 161
Ineson, J. 89
infiltration **15**, 28, 30, 34–5, 75–6,
98–101, 104–5, 127–8, 156
to deserts 104–5
infiltration capacity **15**, 34–5, 98
infra-red imagery 129
infra-red line scanning 129
instability 172–3
interception 14
interface, fresh water/salt water 186, 187,
13.2
gas–liquid 23, 31, *4.2*
interflow **16**, 35, 88, 93, 98, 126–7, *3.1*
intermediate zone **28**, 32, 35, *4.4*
intermittent stream **93**, *8.6*
intrinsic permeability 51
ion exchange 158, 159
isotopes 105, 163–7
radioactive 165–6; stable 165, 166–7
isotropy (-ic) **49**, 50, 55, 63, 88–9, 138,
6.9
Itchen, River 89

jetting 179
juvenile water 157

kanat 84, 108
kelly 121, *9.6*
Khabour, River 92
kinematic viscosity 49, 51, 62
kinetic energy **42**, 43, 44–5, 53, 57–9,
   6.5, 6.7
Kufra oasis 107, *8.12*

laboratory measurement of aquifer
   properties 134–6, 144, 145, *10.2, 10.7*
laminar flow 60–2
landslips 174
lava 77, 86–7
leakage 139
Libyan Desert 105–7, *8.12*
limestone 3–6, 19–20, 29, **79**, 89–92,
   2.1b, c, d
lining (tubes) 114, 117, *9.4*
loading effects 74
Lofthouse Colliery 182
logging of wells and boreholes 130–2,
   145, *10.1, 10.2, 10.7*
London
   decline of potentiometric surface 67,
   170; epidemics 153–4; water supply
   170–1
losing stream **93**, 103, 133
lost circulation 122
lysimeter 127

magma 78
Mammoth Cave, Kentucky 19
manometer **39**, 40, 43–4, 47, 53, *6.2*
Máriotte, Edmé 17
Meinzer, O.E. 65, 68–70, 183
Mendips 3, 167
meniscus 22–4, 31, 23, *4.2, 5.2*
metamorphic rocks 77, **78**, 79, 86, 99,
   181, *8.10*
meteoric water **156–7**, 159, 162–3, 167,
   11.1
Mexico City, subsidence 76
micro-organisms 29, 155
millidarcy 51
mine drainage 162, 181
mine shaft 172, 178–9, 181
models 48, **148**, 149–50
   analogue 149–51, *10.9*; calibration
   150–1; digital 150–1; finite-difference
   150; resistance analogue 48, 149–50,
   10.9; sand tank 148

net gain 171
Neuman S. P. 142
nitrates 158, 160
nuclei (of condensation) 13

oasis 107, 162
   Kufra 107, *8.12*
Ohm's law 48

oil industry 51, 109, 121, 131, 162, 168
oil well 108–9, 121, 122
osmosis 163
overflowing well 12, 67–8, 72, 108, *2.3*
overland flow **15**, 35, 88, 98, *3.1*
oxidising conditions 160
oxygen isotopes 166–7

packers 147, 178, *10.8*
packer testing 147, *10.7, 10.8*
Palissy, Bernard 17
paraquat 9
Paris 46
pathogenic organisms 155–6
Peak District 3
pellicular water 28, *4.4*
Penman, H. L. 33
Penrith Sandstone 83
perched aquifer 66, *7.1*
perched water table 66, *7.1*
perennial head 93, *8.6*
perennial stream 93, *8.6*
permanent wilting point 32
permeability 10–12, 20, 48–51, 55, 59,
   127, 134–6
   anisotropy 49, *6.9*; Chalk 81–2;
   different rock types 77–86; field 51;
   horizontal *6.9*; intrinsic 51; laboratory
   measurement 123, 125, 135–6; Permian
   sandstone *10.7*; reduction 177–8;
   variation with depth 89, *10.7*; vertical
   6.9
permeable 10, 48–9, 50
permeameter 135, 143
Permo-Triassic sandstones 54, 79, **83–4**,
   125, 160, 181, *2.1, 7.6, 10.2, 10.7*
   transmissivity 54; water quality 160
Perrault, Pierre 17–18
Persia 108
petrifying wells 20
physical quality of water 156, 167
piezometers 55, 89, *6.10, 8.1*
piezometric surface 66
plants
   herbaceous 32; role in hydrology 28,
   30–1, 82–3; role in soil formation 29,
   30; wilting 32, 34
plaque 154–5
plaque-forming unit 155
pluvial period 105
poliomyelitis 154
pollutant movement 184
pollution 9, 183–4
pores *see also* voids **6**, 7, 10, 27–8, 31,
   32–3, 78, 81, *2.1*
   in Chalk 81–2, *2.1*; in Permo-Triassic
   sandstones 83, *2.1*
pore-water pressure 72–4, 141, 173–7,
   7.3, 12.3, 12.4
pore-water suction 31–3, 34–5, *5.1*
   measurement of 33
porosity 7, 10, 12, 25–7, 132–6, *2.1, 4.4*
   Chalk 79–83; different rock types

77–87; igneous and metamorphic rocks 86–7; measurement of 132, 134–6; non-aquifers 85; Permo-Triassic sandstones 83, *10.2*; primary **78**, 86; secondary **78**, 86
porous medium 49–50
potential energy **40–1**, 44, 57
potential evapotranspiration **33**, 34, 102
potentiometric surface 12, 53, 55–60, 63, 66, 92, 136–8, 143, *2.3*, *6.10*, *6.13*, *7.2*, *8.4*, *8.5*
decline of in confined aquifers 67, 76; depression of *see* cone of depression
precipitation of calcium carbonate 19–20
precipitation (rainfall) 14, 102, 126–7, *3.1*
pressure **40**, 43–5, 55, 187, *6.7*
changes in aquifers *see* water-level fluctuations; decrease at a constriction *6.7*; difference across a meniscus 23, 24, *4.2*; energy 40–3, *6.2*, *6.5*
pump 55–60, 109–12, 116–17, 169–70, *12.1*
air-lift 111–12, *9.3*; Broad Street 153–4; centrifugal 109, *9.1*; hand 109; piston 109; rotary 109; submersible 111, 117, *9.2*; turbine 111
pump impeller 109–11, *9.1*, *9.2*
pumping station 169, *12.1*
pumping test 136–45, *10.5*
pumping water level 60, *6.13*, *9.2*

qanat *see* kanat
quality of groundwater 107, 145, 153–68, 183
for drinking 153–6; in arid areas 160–2
quickflow 98–9
quicksand 174

radioactivity decay 165, 166, 167
radioactive waste 184
radio-isotope 165–6
radiometric dating 165
rain(fall) 14, 19, 67, 75, 76, 98–101, 104–5, 156–7
acid 157; effective 75
raindrop 14
compacting effect on soil 34
rainfall intensity 35
rainguage 127
Ras-el-Ain spring 92
rating curve 96, *8.8*
recharge **15**, 34, 35, 66–7, 75–6, 102, 104–5, 128, *3.1*
artificial 171
recharge area 67–9, 76, 89, 105, *7.2*, *8.1*, *11.2*
recharge boundary 132, 139, *10.3*
reducing conditions 160
remote sensing 129
reservoir, distribution 170
reservoir, surface 9, 12, 103, 169
residence time 105, 163–4

resistance analogue model 48, 149–50, *10.9*
resistivity surveying 130–1
rest water level 60, *6.13*
reverse osmosis 163
Reynolds number 62
rivers 13, 88, 92, *3.1*
discharge *see* discharge; flow *see* flow
river augmentation 171
rock types as aquifers 77–86
root constant 33–4
rotary table 121, *9.6*
running sand 174
runoff, surface **15**, 98
total 15

sabkha 162
Sahara 104, 162, 165
saline water 161–3, 185–7, *13.2*
salinity 161–3, *13.2*
salt marsh 162, *11.2*
salt water 186–7, *13.2*
salt-water intrusion 186, *13.2*
samples, sampling 115, 123–4, 135–6, 143, *10.4*
sandstone 6, 20, 24, 29, 78–9, 174, *2.1*
Permo-Triassic 54, 79, 83–4, 125, 181, *7.6*, *10.2*, *10.7*; Penrith 83
sanitary protection 114, *9.4*
saturated zone 7, 25, 28, 32, 66, 172, *2.2*, *4.4*
screen **115**, 116–17, 119, 123, *9.4*
sediments 78
consolidated 78–9, 173–5, *9.4*; indurated 78–9; non-indurated 78; unconsolidated 78, 84; well construction in unconsolidated 113, 121, 122, 125, *9.4*
sedimentary rocks 49, 77, **78–9**, 156–7
seepage 90, 107, *8.3*
Selby mine 181
Seine, River 17
semi-permeable membrane 162–3
service reservoir 169–70, *12.1*
sewage 153, 154
shaft, mine 172, 178–9, 181
shaft, well **109**, 123, 139
shear strength 174
sheet piling 177
Snow, Dr John 153
sodium chloride 158–60, 162–3
Soho 153–4
soil 7, 14–15, 28, 29–35
carbon dioxide in 19, 29, 157; formation of 29–30
soil mechanics 173, 175
soil moisture 30
soil-moisture deficit **30**, 33–5, 102–3, 104–5, 128
soil – moisture tension 31
soil water 29, 31
isotopic studies 167
soil – water suction 31

solution *see* dissolution
specific heat 19, 129
specific retention **27**, 28, 30–2, 34, 74, *5.1*
    of chalk 83
specific yield **25**, 27, 28, 30–1, 34, 69, **70**, 71, 74–6, 134, 142, *7.3*
    estimating 75–6; of chalk 81–2; of Permo-Triassic sandstones 83–4
spring 7, 67, **90**, 105, *8.3*
    hot 167; Ras-el-Ain 92
stable isotopes 165, 166–7
staff gauge 95, *8.7*
stage (of pump) 111
stage (of river) **95**, 96–7, *8.7*, *8.8*
stage-discharge relationship **95**, 96, *8.8*; stage hydrograph 95
stalactite 20
stalagmite 20
static water level *see* rest water level
Stephenson, Robert 142
stilling well *8.7*
stomata 31
storage
    changes 74, 82, 98–9, 126–8, *7.5*; elastic 69–70, 141–2; importance of groundwater in drought 102; in conjunctive use 171, *12.2*; release of water from 69–70, 139–42
storage coefficient **71**, 76, 132, 133, 136–8, 142, 150, *7.3*
    measurement of 136–9
stream
    ephemeral **93**, 104; gaining **92**, 93, 101, 103, 133; intermittent 93, *8.6*; losing **93**, 103, 133; perennial **93**, *8.6*
stream channel 35
streamflow 94, 98, 101, 103, 126–7
stress
    effective 174–5, *12.3*; total 174, *12.3*
subsidence 76, 174
subsurface water 7
sulphates 158–60
surface runoff **15–16**, 98
surface tension 21–7, 28, 30, 31, 32, 175, *4.1*, *5.1*

television, borehole 145, *7.7*
temperature of groundwater 156, 167–8
Thames, River 103, 153
    groundwater scheme 170–1; 'leak' 103
Theis, C. V. 138, 139–42, 183
Theis equation 138–42, 148, *10.6*
Third World 187, *13.3*
thundercloud 14
tidal effects on aquifers 74
total runoff 15
tracers 163–4, 166, 184
    dispersion of 184–5
Trafalgar Square wells 67
transmissivity **54**, 55, 60, 63, 76, 116, 132, 136–42, 147–50

of Chalk 54, 81–3; measurement of 136–9
transpiration 14, **15**, 32, *3.1*
travertine 20
tritium 165–6
tunnels 172, 179
turbulence 59, 62–3, 117
turbulent flow 60–2
type curve 138–9
typhoid 154–5

ultrasonic river gauging 96
unconfined aquifer **11**, 54, 57–9, 63, **66**, 68, 69–74, 92, 139–42, *6.10*
    pumping of 139–42
unconsolidated rock 173–4
underflow 127
underground water 1, 3
United Kingdom
    aquifers 79–84, *7.6*; use of groundwater 79
United States Geological Survey 65, 66, 85, 183
unsaturated zone 7–8, 11, 15–16, 28, *2.2*
    of Chalk 83; stress relations 175

vapour, water 13
velocity–area station 95–7
Venturi meter 45
vertical flow 54, 58, 62, 66, 69, 85, 89, 142, 143, *6.10*, *8.1*
vertical hydraulic gradient 55, 57–9, 60, 68, 89, 143, *6.10*, *8.1*
vertical permeability *6.9*
vesicles 77, 86
viruses 154
viscosity 49–50, 51, 62, 63
voids *see also* pores **6**, 7, 10, 28, 77–8, 85, *2.1*

wadis 104, 108
waste, disposal of 184
    hazardous 183; radioactive 184
water as a geological agent 19
    Earth's store of 8, 12, 14; subsurface 3; undergound 1, 3
water balance 12, 33, 126–7, 133
water cycle *3.1*, 12, 14, 16, 34, 88, 166, *3.1*
water distribution *12.1*
water diviner 3
water divining 151–2
water-level fluctuations 72–4, 75, 76, 93, *7.4*, *7.5*, *8.6*
    in Chalk 82
    in Permo-Triassic sandstones 83
water-level recorder *8.7*
water quality 153–68
    effect on aquifer properties 136
water supply 108, 169, 183, *12.1*
water table 7, 8, 10–11, 16, **25**, 28, 32,

35, 54–5, 57, 63, 64, 66, 70, 72, 92–3, 101, 123, 139–42, *2.2, 2.3, 4.3, 4.4, 6.10, 8.5, 8.6*
changes in *see also* water-level fluctuations 74–6, 128; in Chalk 82–3; lowering for excavations 179–80, *12.5*; perched 66, *7.1*
water-table aquifer *see also* unconfined aquifer 66
water witching 151–2
watershed 88
weathering 29, 77, 86
weir 96–7
well 1, 3, 9, 55–60, 108, **109**, 111–23, 169–70, *2.2, 2.3, 12.1*
artesian 55, 68, 72, 108, *7.2*; flow to 54, 55–60, 63–4, 117, 139, *6.11, 6.12*; fully penetrating 55; Joseph's 108; observation 137–8, 143, *10.5*; overflowing 12, 67–8, 72, 108, *2.3*;

petrifying 20; production/pumped 137–9, 141–2, 150, *10.5*
well design 113–18, *9.4*
well development 116
well digging 108–9, 114, 122–3
well discharge 62
well drilling *see* drilling
well efficiency 60, 116
well logging 130–1, 144–5, *10.1, 10.2, 10.7*
well loss 58, 63, 64, 117, *6.13*
well point 180, *12.5*
wilting point 32
World Health Organisation 1

yield-depression curve 59, 63, *6.14*
York 10

zone of fluctuation 34–5, *8.6*